Volume 13

THE GEOGRAPHY
OF WESTERN EUROPE

THE GEOGRAPHY OF WESTERN EUROPE

A Socio-Economic Survey

PAUL L. KNOX

Routledge
Taylor & Francis Group

LONDON AND NEW YORK

First published in 1984 by Croom Helm Ltd

This edition first published in 2016
by Routledge
2 Park Square, Milton Park, Abingdon, Oxon OX14 4RN

and by Routledge
711 Third Avenue, New York, NY 10017

Routledge is an imprint of the Taylor & Francis Group, an informa business

British Library Cataloguing in Publication Data
A catalogue record for this book is available from the British Library

ISBN: 978-1-138-95340-6 (Set)
ISBN: 978-1-315-65887-2 (Set) (ebk)
ISBN: 978-1-138-95534-9 (Volume 13) (hbk)
ISBN: 978-1-315-66640-2 (Volume 13) (ebk)

Publisher's Note
The publisher has gone to great lengths to ensure the quality of this reprint but points out that some imperfections in the original copies may be apparent.

Disclaimer
The publisher has made every effort to trace copyright holders and would welcome correspondence from those they have been unable to trace.

The Geography of Western Europe

A SOCIO-ECONOMIC SURVEY

PAUL L. KNOX

CROOM HELM
London & Sydney

BARNES & NOBLE
Totowa, New Jersey

© 1984 P.L. Knox
Croom Helm Ltd, Provident House, Burrell Row,
Beckenham, Kent BR3 1AT
Croom Helm Australia Pty Ltd, First Floor, 139 King Street,
Sydney, NSW 2001, Australia

British Library Cataloguing in Publication Data

Knox, Paul L.
     The geography of Western Europe. – (Croom
     Helm series in geography and environment)
     1. Anthropo-geography – Europe 2. Europe –
     Social conditions
     I. Title
     940.55'8   GF540
     ISBN 0-7099-1523-3
     ISBN 0-7099-1524-1 Pbk

First published in the USA 1984 by
Barnes & Noble Books,
81 Adams Drive,
Totowa, New Jersey, 07512

Library of Congress Cataloging in Publication Data

Knox, Paul L.
     The geography of western Europe.

     Includes index.
     1. Europe – Social conditions – 20th century.
2. Europe – Economic conditions – 20th century. 3. Anthropo
-geography – Europe. I. Title.
HN375.5.K55   1984   306'.094   84-12330
ISBN 0-389-20512-5

Typeset by Columns of Reading
Printed and bound in Great Britain

CONTENTS

TABLES

FIGURES

ACKNOWLEDGEMENTS

I should like to thank my colleagues Jimmy Caird, Nick Ford and Huw Jones and my wife Lynne for their comments on draft sections of the book. The data from the Eurobarometer series were made available by the Inter-University Consortium for Political and Social Research (USA). These data were originally collected by Jaques-Rene Rabier, Special Adviser to the Commission of the European Communities, and by Ronald Inglehart of the University of Michigan. Neither the original collectors of the data nor the Consortium bear any responsibility for the analyses or interpretations presented here.

I am also grateful to the following for permission to reproduce copyright material:

Cambridge University Press for Figure 5.5, based on Figure 6, p. 428 of *Regional Studies*, vol. 16, 1982; the Commission of the European Communities for Figure 5.9, based on an illustration in *The Community and Its Regions*, 1980; the Council of Europe for Table 2.2, based on Table 6.1, p. 145 in *Demographic and Social Change in Europe*, 1975-2000, by M. Kirk, Council of Europe, 1981; Elsevier Scientific Publishing for Table 7.2, based on Table 1, p. 7 of the *European Journal of Political Research*, vol. 11, 1983; Professor P. Hall for Figure 6.1 and Table 6.1, based on Figure 4.16, p. 156 and Table 4.9, p. 155 in *Growth Centres in the European Urban System*, Heinemann Educational Books, 1980; the International Labour Office for Table 2.3, based on Table 2, p. 5 in *Return Migration from West European to Mediterranean Countries*, by H. Entzinger, ILO, 1978; Professor L.H. Klaasen for Table 6.2, based on Table 9.7, p. 132 in *Dynamics of Urban Development*, Gower, 1981; Dr H. Leitner for Figure 6.6, slightly modified from Figure 2, pp. 98-9 in *Mitteilungen der Osterreichischen Geographische Gesellschaft*, vol. 123, 1981; Martinus Nijhoff Publishing for Figure 5.7, slightly modified from Figure 8, p. 109 of *Spatial Inequalities and Regional Development*, edited by H. Folmer and J. Oosterhaven, 1979; Dr M. Pacione for Figure 6.4, slightly modified from Figure 6.3, p. 232 in *Urban Problems and Planning in the Developed World*, Croom Helm, 1981; Pergamon Press for Figure 6.3, based on Figure 6, p. 61 of *Geoforum*, vol. 4, 1970; the Population

Council for Figure 2.5, slightly modified from Figure 9, p. 41 of *Population and Development Review*, vol. 7, 1981.

P.L. Knox
Dundee

1 INTRODUCTION

The geography of Europe has traditionally been delineated in terms of its physical geography, its ethnography and culture and its economic specialisations, while Europe itself has long been justified as a meaningful region because of the predominance of Caucasian stock, Indo-European languages and Christianity, the intensive organisation of its agriculture, the extensive development of its urban system, its high degree of industrialisation and its correspondingly high levels of material well-being (Gottmann 1969; Mead 1982). Within this context geographical analysis has typically been organised around the physical and/or economic landscapes of the major sub-regions of individual countries or groups of countries. This is the Europe of the Alps, the coalfields, the Rhine Rift Valley, the Spanish Tableland and the North Italian Plain, described in detail by geographers from every European country (Marchant 1970). But such an approach has become increasingly inadequate. For one thing, the conceptual and theoretical development of geography has come to emphasise systematic approaches at the expense of regional synthesis. Hence the emergence of systematic analyses of the geography of Europe such as those by Jordan (1973) and Mellor and Smith (1979). At the same time, the postwar division of Europe into two distinctive geopolitical units has overwritten the cultural and physiographic unity of the continent, making it increasingly difficult to justify as a single coherent region, at least in terms of human geography. Hence the emergence of systematic analyses dealing separately with Western Europe (Diem 1979; Ilbery 1981). The actual geography of Western Europe, meanwhile, has undergone significant changes as a result of demographic trends, changes in patterns of urbanisation, the reorganisation of political and administrative frameworks, the creation, enlargement and reorganisation of the European Community, the expansion of welfare states and, above all, structural economic change. Hence this book.

As the title suggests, the book is oriented towards an understanding of the *social* as well as the economic geography of Western Europe. Here again a word of explanation is appropriate. The term 'social geography' is used here in its original sense (Vallaux 1908) as an alternative to 'human geography'. This is not simply a matter of semantics, but rather an attempt to encapsulate a somewhat differently composed

1

spectrum of phenomena than has come to be associated with the term human geography. This, in turn, is another consequence of conceptual and theoretical developments within geography. In short, the emphasis has shifted away from the likes of ethnography, cultural geography, commercial geography and abstract economic models towards political economy, welfare geography and humanistic geography. It is important, of course, not to throw the baby out with the bath water. Many of the features of pre-1950 Europe which were the focus of 'traditional' human geography remain as distinctive components of present-day Europe: traditional crop regions, field systems, peasant communities, village forms, townscapes, and the territorial mosaic of religious and linguistic groupings, for example. But they must be seen in conjunction with new features which have emerged in response to demographic, economic and social transition. The objective of this book is to review the salient features of the social and economic geography of Western Europe, with particular emphasis on the nature and outcome of post-war changes. The book is organised systematically, beginning with a brief introduction, in this chapter, to the nature and extent of socio-economic differentiation within Western Europe.

Dimensions of Differentiation and Change

The socio-economic profiles of Western European nations reflect both the unity and the diversity of the region as a whole (Table 1.1). Overall, the picture is one of relatively high levels of prosperity, with national economies continuing to expand at a respectable rate. Most economies no longer rely much on agriculture but on manufacturing and, increasingly, service industries. Birth-rates are generally relatively low, while the expectation of life is high and infant mortality-rates are low. Substantial numbers of people receive an extended formal education, and the ownership of consumer goods such as television sets is high. Furthermore, the provision of collective facilities such as hospital beds is generally high, largely because of the high proportion of national incomes allocated to social welfare programmes. Differentials between nations remain substantial, however. Gross Domestic Product (GDP) *per capita* ranges from 1,750 European Units of Account (EUA) in Portugal to 11,451 EUA in Switzerland; the infant mortality rate ranges from 6.9 per 1,000 live births in Sweden to 26.0 in Portugal; while in Norway there are more than twice as many hospital beds per inhabitant than in Greece, Portugal and Spain. As Almeida (1980) notes, many of

Table 1.1: National Socio-economic Profiles, *c.* 1980

	Population (000s)	Crude birth-rate per 1,000 inhabitants	Infant mortality-rate per 1,000 live births	Expectation of life at birth (males)	GDP per capita^a	% change in GDP per capita 1970-80	% labour force in agriculture	% labour force in service sector	% labour force unemployed	% of population with tertiary education	TV sets per 100 population	Population per hospital bed	Social security expenditure as % of GDP
Austria	7,505	12.5	14.3	68.9	7,368	3.7	10.1	49.2	1.9	*	24	88	21.4
Belgium	9,859	12.7	11.0	68.6	8,530	3.2	3.0	63.6	9.4	8.9	29	111	27.7
Denmark	5,123	11.2	8.4	71.2	9,310	2.3	8.1	64.4	6.2	10.9	32	103	28.0
Finland	4,779	18.7	7.6	67.4	7,491	3.5	11.6	54.0	5.3	*	36	65	19.9
France	53,714	14.7	10.0	69.2	8,750	3.6	8.8	56.2	6.4	9.1	37	97	25.8
Greece	9,599	15.6	18.0	70.1	3,000	4.8	30.3	39.5	1.1	6.5	13	156	12.0
Ireland	3,401	21.9	11.2	68.8	3,793	4.1	19.2	48.4	8.3	5.1	21	94	22.0
Italy	57,070	11.3	14.3	69.0	4,995	3.2	14.2	49.3	8.0	9.6	22	94	22.8
Luxembourg	365	11.4	11.5	67.3	9,041	3.1	5.7	57.0	0.7	5.1	29	85	26.5
Netherlands	14,150	12.8	8.6	71.5	8,600	2.9	4.6	64.8	4.8	8.9	29	99	30.7
Norway	4,086	15.5	11.9	72.3	10,058	4.6	8.5	61.7	1.7	8.7	27	68	20.1
Portugal	9,884	16.3	26.0	65.1	1,750	4.8	28.5	36.8	7.8	5.0	8	187	12.0
Spain	37,340	15.1	11.1	69.7	4,070	3.8	18.9	45.0	12.7	7.9	19	190	14.0
Sweden	8,311	11.7	6.9	72.4	10,612	1.9	5.6	63.1	2.0	13.5	37	66	34.0
Switzerland	6,366	11.3	9.0	70.2	11,451	1.2	6.4	53.3	0.2	*	29	87	16.1
West Germany	61,566	10.1	12.6	68.6	9,529	2.9	6.0	50.0	3.4	10.3	31	84	28.3
United Kingdom	56,010	13.5	11.9	69.6	6,727	1.9	2.6	60.9	6.9	4.7	32	117	21.4
USA	227,658	15.8	12.6	69.9	8,199	3.0	3.6	66.4	6.0	18.0	57	155	14.2

* data not available

Note: a. In European Units of Account.

Source: National statistical abstracts.

the differentials between nations support the idea of a distinction between a prosperous regional 'core' (Austria, Belgium, Denmark, France, Luxembourg, the Netherlands, Norway, Sweden, Switzerland and West Germany) and a relatively poor 'periphery' (Greece, Ireland, Portugal, Spain). But such a division, even allowing for a 'semi-peripheral' category (Finland, Italy, the United Kingdom), is by no means appropriate to every aspect of economic, demographic and social life: Ireland and Italy, for example, have a better ratio of population to hospital beds than Belgium and Denmark, while the economies of Spain and Portugal have been growing at a much faster rate than those of Sweden and West Germany.

Table 1.2: Value Systems in Selected European Countries

	Materialists	Intermediate	Post-materialists
France	30	51	19
Denmark	26	61	13
Belgium	22	66	12
Netherlands	33	58	9
Luxembourg	27	64	9
Italy	40	53	7
Ireland	40	56	4
West Germany	52	44	4
Great Britain	48	48	4
Northern Ireland	60	38	2

Source of data: Morrison (1977)

Meanwhile, socio-economic change has brought with it some important changes in the social values and attitudes of postwar Europe. Inglehart (1971) has pointed to the emergence of a 'post-materialist' element in West European society as a result of increasing affluence coupled with the development of education and the influence of the mass media since 1950. These post-materialists are oriented towards democratic participation, self-management, rights for ethnic and cultural minorities, environmental protection, the expansion of progressive social welfare programmes, and so on, often pursuing their objectives outside the traditional channels of political participation. 'Materialists', on the other hand, are oriented towards rapid economic growth, higher levels of consumption, limitations on public expenditure (except on defence and law and order) and the maintenance of social stability. In a recent survey of people's values in nine countries, Morrison (1977) found that materialists outnumbered post-materialists by more than five to one, though the value system of exactly half of the 8,400 respondents was found to be intermediate, with no clear disposition towards either materialism or post-materialism. As Table 1.2 shows, there are considerable differences between countries in the balance of these value systems. West Germany and the United Kingdom are clearly the most materialistic, with only a small element of post-materialists. By contrast, materialism is much less prominent in Belgium, Denmark, Luxembourg and France, where post-industrial values are held by a sizeable proportion of the population. The details of these variations need not detain us here. Indeed, it remains to be seen whether or not the post-materialist value system will emerge unscathed from the economic recession of the late 1970s and early 1980s. What is important is that we should recognise the significance of people's feelings and attitudes to an understanding of the social geography of Western Europe. The question of materialist versus post-materialist value systems can thus be seen not only as a reflection of people's changing circumstances, but also as a determinant of their social organisation and, consequently, of patterns of political conflict, resource allocation, and so on. Moreover, people's values and attitudes are themselves an important facet of the cultural variations which characterise the social geography of Western Europe; and they are not always straightforward reflections of the traditional cleavages of race, language, religion and class. Take, for example, the question of women's status. Although the geography of Catholicism is a significant overall correlate of national attitudes towards women's problems in contemporary society, it is by no means the only discriminant. Rather, there appear to be several distinctive orientations to the issue (ranging from militance through indifference to support for the *status quo*), each of which, though varying in popularity, tends to be represented in most countries, irrespective of religious adherence (Commission of the European Communities 1975).

It should be emphasised that, although both the unity and the diversity of Western Europe are apparent at the level of nation states, and although the international context is important to establish for many purposes, it is *regional* differentiation that is the focus of this book. The reason is simply that many of the dominant features of spatial differentiation within Western Europe tend to cut right across national boundaries. Figure 1.1 illustrates some of these, including the localisation of industrial decline in parts of Belgium, France, West Germany and the United Kingdom, the persistence of peasant-based agriculture in parts of France and the Iberian peninsula, the emergence of tourist-dependent sub-regions in some of the more physically and/or climatically attractive parts of peripheral Europe, and the differential effects of the dynamics of urbanisation which have resulted in the relative stagnation of cities like Cardiff and Glasgow while the likes of Barcelona, Lisbon and Naples have experienced an intensification of urban development. As we shall see in subsequent chapters, these distinctive features of contemporary Europe represent the imprint of a variety of overlapping, intersecting and interacting forces, all of which are conditioned by the legacy of earlier phases of national and regional development.

We begin, in Chapter 2, by establishing the demographic background to Western Europe. As Hooson (1960) and others have argued, the distribution of population can be considered as the 'essential expression' of geography. The location and density of population, for example, reflect the overall pattern of resources and economic development, while patterns of mortality provide a useful index of regional variations in social well-being. Similarly, migration streams are closely related to patterns of regional economic change, and trends in fertility are sensitively attuned to certain aspects of social change. Meanwhile, all these components of population change exert important reciprocal effects on regional social geography, shifting the balance of supply and demand in local labour markets and housing markets, and altering the pattern of need for social services of all kinds.

In Chapter 3 attention is turned to the political quilt and the changing relationships between ethnography, territorial organisation and the political economy of Western Europe. The system of nation states has been of fundamental importance in shaping patterns of economic, political and social development within Western Europe. On the one hand they helped to foster the political economy of industrial capitalism; on the other, they have fostered xenophobia and jingoism, and have contributed to the subordination of minority cultures and ethnic groups. Meanwhile, a 'New Europe' has developed around the framework of supranational organisations like the European Community – partly in response to the international political climate in the aftermath of the Second World War, and partly in response to economic imperatives stemming from the internationalisation of capital. Within this framework run the counter-trends of devolution and separatism, along with a variety of political cleavages – class-based, religion-based, urban-rural – which in turn are articulated through different types of party and electoral systems.

Chapter 4 deals with the rural dimension of Western Europe's social and economic geography. Although the relative importance of rural population and employment has greatly decreased, rural regions still contribute a good deal to the character of regional landscapes, politics, settlement patterns, and patterns of social well-being. Contemporary rural Europe still retains much of the legacy of past forms of economic and social organisation. The geography of agriculture is of course closely conditioned by the physical environment and by the different systems of farming and landholding which have evolved in different parts of Europe, while many areas have retained 'traditional' settlement patterns and a peasant culture. But most of rural Europe has been transformed over the postwar period as new dimensions have been added to its geography. The structural reorganisation of agriculture has created new kinds of land-use conflict, reinforced the process of rural depopulation, and brought into question the viability of many smaller communities. State intervention in agriculture and rural affairs, meanwhile, has come to play an increasing role in recasting and redirecting the geography of rural regions. In addition, the general 'revolution' in personal transport has opened up many rural regions to tourists, retired households, commuters and 'neo-rural' colonists, thus creating new social milieux within rural communities.

The industrial regions of Western Europe have also been transformed over the postwar period as the changing imperatives of industrial technology and economic organisation have left their imprint on local labour markets. Chapter 5 describes the impact of these changes against the background of the patterns of regional industrial development which had been the product of earlier phases, waves and cycles of economic development. The most fundamental aspect of change during the postwar period has been the general shift in emphasis away from manufacturing industry to tertiary and quaternary activities. This has been accompanied by changes in the size and spatial organisation of business enterprises and by a massive increase in public sector employment and in government intervention in economic affairs. At the regional level, these changes have been reflected by changes in the pattern of unemployment, of female labour, of wage-rates and of employment opportunities. The net result is often interpreted in terms of *centres* of economic power and social well-being and *peripheries* of limited (or suppressed) potential for economic development and a high incidence of disadvantaged households.

The urban dimensions of Western Europe are examined in Chapter 6. Once again, the legacy of past phases of development is emphasised, with attention being given both to the system of towns and cities and to their ecological structure. As in other highly urbanised and 'advanced' regions, recent urban development has been dominated by metropolitan decentralisation. This is reflected most clearly by the relative (and, in some cases, absolute) decline of the metropolitan cores of northwestern Europe and the expansion of their suburban and satellite communities. Meanwhile, urban development in southern Europe has intensified as migrants continue to arrive from the countryside, more than compensating for the departure of many middle-class households to expanding suburbs and nearby metropolitan villages. Indeed, the European league table of urban growth is dominated by southern cities, including Madrid, Santander, Turin, Milan, Rome and Barcelona. Many of the problems currently experienced by European cities are related to these changes in the broad pattern of urban development. Thus the rapidly growing cities of southern Europe are characterised by acute problems of infrastructural provision and the overloading of public services throughout metropolitan areas. The metropolitan regions of the industrial northwest, on the other hand, are characterised by inner-city problems in which residual, service-dependent households figure prominently. It is in these cities that municipal socialism has developed most fully, though the whole of Western Europe has a long history of urban planning and municipal control. Indeed, town planning has not only been a major product of European urbanisation, it has also been an important factor in reshaping and dictating the character of European cities themselves.

In a similar way, the welfare state has been a major product of European socio-economic development, and has also been an important factor in influencing patterns of social well-being within and between nations. In Chapter 7 the performance of Europe's welfare states is considered against the background of the current extent of regional inequality and of the evidence for convergence in relative levels of regional well-being. The persistence of social and economic inequality throughout Western Europe is not surprising, given that the bulk of the massive postwar expansion in social welfare programmes has in fact been aimed at 'average' households rather than low-income or disadvantaged groups. But in addition to the consequent gaps in welfare provision, the expansion of most welfare states has occasionally and unintentionally created *new* forms of deprivation and disadvantage. Moreover, the widespread economic recession which began in the early 1970s accentuated the vulnerability of more and more people while making it increasingly difficult – both economically and politically – to finance existing social welfare programmes. As a result, it has been suggested that Western Europe's welfare states have entered a new stage of reformulation and retrenchment; though, as we shall see, individual countries have responded in different ways. Similarly, the 'resource squeeze' has affected some local government areas much more than others, thus adding a new dimension to the geography of the 'social wage' which has become an important component of the overall social well-being of communities and regions.

Figure 1.1: Some Aspects of Spatial Differentiation in Postwar Europe

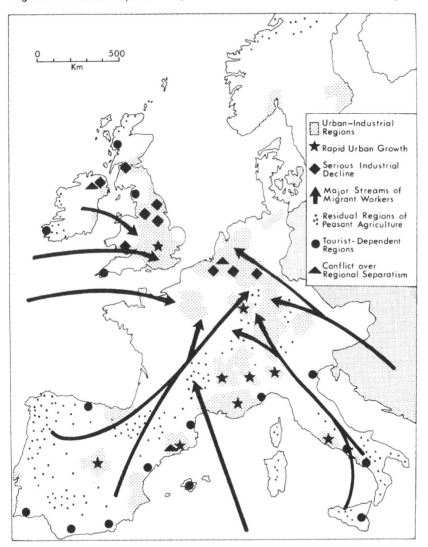

2 THE DEMOGRAPHIC BACKGROUND

In Western Europe, as in other regions, the geography of population and the dynamics of population change are closely interrelated with processes of economic development and social change. Patterns of density, fertility, mortality and migration are often a direct reflection of economic, social and political conditions; they are also important determinants of economic change and social well-being at both national and regional levels. Although it is not always easy to unravel cause and effect, it is important to set out at least the major demographic patterns and trends within the region and to identify their linkages with socio-economic development.

Population Distribution and Density

A distinctive characteristic of Western Europe as a whole is the size and relative density of its population. With only 2 per cent of the world's land surface, Western Europe contains 8.6 per cent of its population at an overall density of nearly 100 persons per km^2. By comparison, the United States contains only 5 per cent of the world's population at a density of 24 persons per km^2. Within Western Europe the highest national densities (344 per km^2 in the Netherlands, 325 per km^2 in England and Wales, 322 per km^2 in Belgium, and 247 per km^2 in West Germany) match those of Asian countries such as Japan (311 per km^2), the Republic of Korea (382 per km^2) and Sri Lanka (225 per km^2). On the other hand, population density in Finland, Norway and Sweden stands at about 15 persons per km^2: the same as in Kansas and Oklahoma.

This points to a second distinctive feature of the human geography of Western Europe: the existence of a densely populated core and a sparsely populated periphery. Broadly speaking, this core can be regarded as stretching from the United Kingdom through Belgium, the Netherlands, West Germany and Switzerland to Italy: the six countries with the highest densities and, moreover, with relatively even internal distributions of population at the provincial level (Arnold, Danieli and Zacchia 1981; Biraben and Duhourcau 1973). At the more detailed level of resolution shown by Figure 2.1 it is clear that this core is itself dominated by an intermittent zone of extremely high densities (i.e. more than 250 persons per km^2). This zone can be traced from its northernmost outlier in the central lowlands of Scotland through the industrial heartlands of the Tyneside, Merseyside, Manchester, West Yorkshire, South Yorkshire and West Midlands conurbations to metropolitan southeastern England and across the Channel to industrialised northeastern France, Belgium and the Randstad Holland conurbation. Southeastwards, it extends into the agricultural-industrial environment of the *Börde* country between Aachen and Hannover and thence to the heavily industrialised Ruhr conurbation. High densities continue unbroken to the south, following the Rhine as far as Karlsruhe and extending westwards to the Saar coalfield and eastwards to the Neckar basin and Stuttgart. Thereafter, the southward extension of the high-density zone can be traced only intermittently, appearing in the Swiss *Mitteland* and again in northern Italy around Turin, Milan and Venice and their rich, commercially farmed hinterlands. Beyond this core zone regions of comparable density are mostly restricted to those containing major metropolitan centres: Bremen and Hamburg in West Germany; Lyon, Marseille and Paris in France; Barcelona, Bilbao and Madrid in Spain; Lisbon and Porto in Portugal; Dublin in Ireland; Vienna in Austria; Geneva in Switzerland; and Genoa, Naples and Rome in Italy. The exceptions are the poor but densely settled rural areas in the heel of Italy and the coastal lowlands of Sicily.

Most of the rest of Western Europe to the east of France and south of Sweden is peopled at intermediate densities. In France extensive areas of rural upland are reflected in densities of less than 50 persons per km^2, with parts of the Massif Central and the southern Alps containing less than 25 persons per km^2. Similarly, most of the relatively inhospitable Iberian meseta and the dry uplands which dominate Greece are sparsely populated, as are the upland areas of Britain. Western Europe's real 'empty areas', however, are the vast interior regions of Scandinavia. Much of Iceland and the northern regions of Norway, Sweden and Finland are virtually uninhabited save for a few small mining and industrial towns and a scatter of small fishing ports.

Demographic Trends and Social Well-being

Western Europe's population is currently characterised by a very slow rate of natural increase, compared with other world regions. While the world population as a whole increased by 17.8 per cent between 1970

and 1980, the population of Western Europe increased by only 4.7 per
cent, from 356 million to 373 million. This figure of course hides
significant national and regional variations, though only Greece, Iceland,
Ireland, Portugal and Spain were growing naturally by more than 4
persons per 1,000 in 1980. Austria, Belgium, Denmark, Sweden and the
United Kingdom were virtually static, while natural change in West
Germany was clearly negative, at around −3 per 1,000.

Figure 2.1: Population Density in Western Europe *c*. 1980

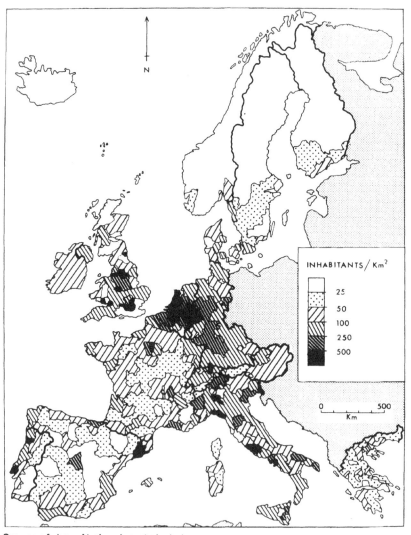

Source of data: National statistical abstracts.

The Historical Demography of Western Europe

These relatively low rates of natural increase have been characteristic of Western Europe for several decades. They are the product of a series of transformations in vital rates which began in the eighteenth century, and which have broadly followed the classic pattern of demographic transition accompanying the evolution from pre-industrial to post-industrial society. The demographic transition model does not precisely describe events in any one country, but it was founded on European experience and has been generally accepted as a useful framework within which to interpret long-term demographic change. The basic model consists of four episodes:

(1) An initial phase of high birth-rates and high death-rates, resulting in low rates of natural increase. Death-rates fluctuate violently in response to disease and economic catastrophes, causing occasional 'demographic crises' of population decline.
(2) A phase of falling death-rates while birth-rates continue at a high level. The margin of births over deaths thus increases considerably, producing a 'demographic explosion'.
(3) A phase during which the death-rate, which cannot of course decline indefinitely, begins to stabilise at a relatively low level while birth-rates begin to decline at a significant pace. The result is that the rate of growth of the population begins to slow up as the margin of the excess of births over deaths declines.
(4) The 'late' phase, when both birth and death-rates are at low levels, with the former marginally higher than the latter; a low rate of increase therefore results. Death-rates remain fairly stable, but birth-rates fluctuate moderately over the short term.

Economic development is seen as the moving force behind this sequence of events. In short, economic development brings lower mortality, and lower mortality helps to induce lower fertility: it is no longer necessary to have six children in order that two or three will survive, while economic development brings compulsory education and laws against child labour, converting children from economic assets to liabilities. In detail, however, the interdependent processes of economic development, social 'modernisation' and population change have been complex, subtle and diverse. Indeed, demographers continue to engage in a lively debate as to the mechanisms and underlying causes of the demographic changes which have taken place (van de Walle and Knodel 1980).

Explanations of Mortality Decline. It is widely accepted that the decrease in mortality which precipitated the secular growth in population in the eighteenth century is largely attributable to a decline in deaths from infectious disease. Beyond this, however, there is considerable disagreement as to the reasons why such deaths should have decreased. The major contending hypotheses are as follows:

(1) Advances in medical science and an expansion of medical facilities were crucial in accounting for the decline in mortality from infectious disease. Particular emphasis is placed on the expansion of hospitals, the setting up of dispensaries for the poor, the introduction of smallpox inoculation (compulsory in a number of countries), and on national quarantine measures designed to reduce the incidence of typhus and dysentery (see, for example, Griffith 1926; Razzell 1977).

(2) Improved standards of personal hygiene – especially the increased use of soap and washable cotton underwear – facilitated a fall in mortality by reducing the danger of infection (see, for example, Razzell 1974).

(3) Improved public health developments – especially water purification and sewage disposal – led to marked decreases in mortality from intestinal disease (cholera, dysentery, typhoid).

(4) The introduction of new crops – particularly the potato – and the structural reform of the agricultural sector brought about an improvement in diet which in turn improved people's resistance to disease (see, for example, Langer 1975; McKeown, Brown and Record 1972).

(5) Changes in the virulence of specific diseases and in the latent immunity of the human population brought about a spontaneous decline in mortality.

The real issue, of course, is the *relative* importance of each of these factors in different parts of Europe and at different times. The 'medical' hypotheses (1-3) are generally credited with accounting for only a small proportion, at best, of the decline in mortality. Widespread improvements in public health and hygiene were not evident until the second half of the nineteenth century; and the major breakthroughs in medical science all occurred well after the initial decrease in mortality. Moreover, it is clear from a number of case studies that the introduction of smallpox inoculation did not make a very big impact on existing levels of mortality, while the effect of quarantine measures has proved difficult to quantify and so remains in doubt (Lee 1979a).

The 'Malthusian' hypothesis (4), argued insistently by McKeown *et al.* (1972) and rejected just as forcefully by Razzell (1974), finds support in some case studies but not in others. In Denmark, for example, changes in the structure of land tenure between 1770 and 1800 were accompanied by a doubling of grain production which evidently contributed to a significant fall in mortality (Andersen 1979). On the other hand, Drake's (1969) assertion that the introduction of the potato was of major demographic significance in Norway has recently been put in doubt by the calculations of Lunden (1976), who attributes a maximum reduction of 13 per cent of the decrease in mortality to the nutritional impact of the potato. Firm evidence relating to the 'autonomous decline' hypothesis (5) is also hard to come by, though it is generally agreed that there occurred a spontaneous change in the virulence of scarlet fever in the nineteenth century. It has also been suggested that such a change may have *contributed* to the decline in mortality from tuberculosis, typhus and cholera (McKeown and Record 1962).

Explanations of Fertility Decline. It is now recognised that the change from high levels of fertility to closely controlled fertility at lower levels did not occur simply as a consequence of industrialisation and urbanisation. Western Europe had experienced a widespread fall in fertility well before the onset of the industrial revolution, largely as a result of a distinctive pattern of late marriage and widespread celibacy (Hajnal 1965). This is popularly attributed to constraints of land availability in peasant societies. These constraints, it is argued, forced many individuals to delay marriage until the death of the patriarch and the consequent release of the land necessary to provide a living for a family (see, for example, Berkner and Mendels 1978). In detail, however, the relationships between nuptiality, fertility and rural resources are complex, often providing conflicting evidence (Braun 1978; Gaunt 1976; Knodel and van de Walle 1979). In overall terms, it seems that pre-industrial Western Europe had developed a 'homeostatic' demographic regime in which procreation was roughly matched to the carrying capacity of the environment by a variety of mechanisms (Smith 1977).

With the onset of the industrial revolution there occurred a further and more dramatic decline in fertility. Again, however, it is important to stress that the relationship between economic development and fertility decline was by no means simple and straightforward. The general fall in the birth-rate cut across national boundaries and regional economic differences; although a significant degree of economic backwardness, as in the case of Portugal (Morgado 1979), southern Italy (Del Panta 1979), and northern Sweden (Hofsten and Lundström 1978) did tend to retard the general process. Moreover, the transition to lower levels of fertility took place over a very limited period of time. Within Germany, for example, the first significant fall in fertility occurred in the 1870s, with the last region undergoing an equivalent transition on the eve of the First World War (Knodel 1974). The same phenomenon occurred in other countries: once the decline had been initiated, the adoption of the new pattern of fertility was quickly assimilated throughout the country (Andersen 1979; Coale 1969; Deprez 1979). Not surprisingly, it has proved impossible to determine anything like a threshold of economic development or urbanisation at which a significant fall in fertility was precipitated (see, for example, van de Walle and Knodel 1967).

Nevertheless, it is not disputed that the industrial revolution was the underlying cause of modern fertility decline in Western Europe. Improvements in science, technology and in *per capita* incomes, for example, brought about a significant decline in infant mortality, which undoubtedly played a decisive role in reducing the birth-rate. Just as important were the complex socio-cultural changes which characterised the urban-industrial transition. The changes in social organisation, the role of the family, education, religious adherence and so on which were triggered by the imperatives of industrial capitalism all interacted in such a way as to encourage smaller families. Moreover, the very process of urbanisation facilitated the rapid diffusion of changes in norms and values relating to child-bearing and parenthood, thus explaining the way in which the fall in birth-rates cut across national boundaries and regional economic differences. The ways in which the socio-cultural changes brought about by industrialisation and urbanism encouraged fewer births were several:

(1) The division of labour necessary to sustain factory production destroyed the economic logic of extended family systems; without the support of such a system, child-rearing incurred much higher economic, physical and emotional costs.
(2) The growth of formal education required by industrialisation undermined religious adherence; this, in turn, weakened the effectiveness of traditional religious exhortations to reproduce.
(3) The division of labour and the growth of formal education, reinforced by legal restrictions on child labour, also reduced the economic value of children to parents, since children could no longer be called upon to contribute to family responsibilities.
(4) The change from ascribed status to achieved status under the new

class structure forged by industrialisation meant that income was increasingly directed towards the symbols and activities of social rank, thus leaving proportionally less with which to sustain a large family.

(5) The increasing division of labour and the expansion of clerical jobs drew more women into the labour force. This not only had the immediate effect of reducing their willingness and capacity to raise children, but also contributed to the long-term emancipation of women from their traditional role as child-bearing domestic servants (Lesthaeghe 1983).

Population Growth and Economic Development. The relationship between economic and demographic change in Western Europe was by no means a one-way affair, however. The rise in European population from 1750 onwards had important reciprocal relationships with economic development, both in terms of the supply of labour and in terms of aggregate demand. The decline in mortality led to a very rapid growth in both the number and the proportion of economically active persons. In Sweden, for example, the age group 20-25 expanded by over 50 per cent during the early nineteenth century; and in Prussia the age group 20-39 expanded by 47 per cent between 1816 and 1840. As Lee (1979b) observes:

> This sudden increase in the size of the available labour force, accompanied by an equally dramatic reduction in the dependency ratio, facilitated continuous economic development in a number of European economies through ensuring an increased supply of labour at zero or marginal opportunity costs. (p. 23)

Such growth created its own independent momentum, with the initial growth of population resulting in a subsequent wave of births, and these in turn producing a further wave some 20 years later; and so on. Furthermore, each boost in population, accompanied by a corresponding rise in the number of marriages (though not necessarily in the marriage-rate itself), served to stimulate economic growth by increasing the level of demand for such items as houses and consumer goods (Easterlin 1966). Lee (1979b) has speculated that these demographic 'shock waves' may have influenced the emergence of national and international business cycles. Another important consequence of the initial population expansion occurred in several countries when the agricultural sector was no longer able to absorb the supply of local labour. This led to an expansion of 'proto-industrialisation' involving craft production and, in particular, textile manufacture. Such developments were important in Belgium, the Netherlands and parts of Germany and Switzerland in fostering a local entrepreneurial class and in facilitating the process of capital accumulation in rural areas. A second and more widespread response to rural overpopulation was of course migration, which is discussed in more detail below (p. 31).

Recent Change and Present Patterns

Mortality. Most Western European countries have experienced low and relatively stable levels of mortality, by world standards, since the 1930s (Caselli and Egidi 1981). In Switzerland, for example, the crude death-rate levelled off at around 12 deaths per 1,000 population after 1925, and has since shown only a small improvement (to 9 per 1,000 in 1980); in Spain, the same levelling-off took place after 1945. Nevertheless, significant differences persist in the gradient of mortality at the regional level (see, for example, Myklebost 1981; Pringle 1982). A recent survey of variations in life expectancy at birth showed that there is a difference between the best and worst-off regions within Western Europe of over ten years (van Poppel 1981). Moreover, comparisons with the particularly favourable mortality experience of Sweden showed that some countries were lagging by the equivalent of 30 or 40 years. Portuguese males born in the 1970s, for example, had the same life expectancy as Swedish males born in the 1930s, while Austrian males born in the 1970s could expect to live only as long as Swedish males born around 1940. Such differentials are of course influenced by national and regional variations in class and occupational structure but, as van Poppel points out, much of the regional variation in mortality must be attributed to local socio-economic factors, including social conditions, environmental quality, and the quality and availability of medical care facilities. One index of mortality which is particularly sensitive to such factors is the infant mortality-rate, usually expressed as the number of deaths under one year of age per 1,000 live births. Indeed, infant mortality is widely regarded as one of the most useful yardsticks of spatial variations in social well-being. It not only reflects the immediate living conditions experienced by pregnant women and infants, but has also been found to be closely correlated with community levels of poverty, with morbidity from infectious disease, and with social instability (Knox 1982a). It also has the advantage of being based on reliable data collected on a regular basis in every Western European country (Höhn 1981). It is, therefore, useful to examine in some detail the nature and extent of regional variations in infant deaths.

Figure 2.2: Infant Mortality in Western Europe *c*. 1980

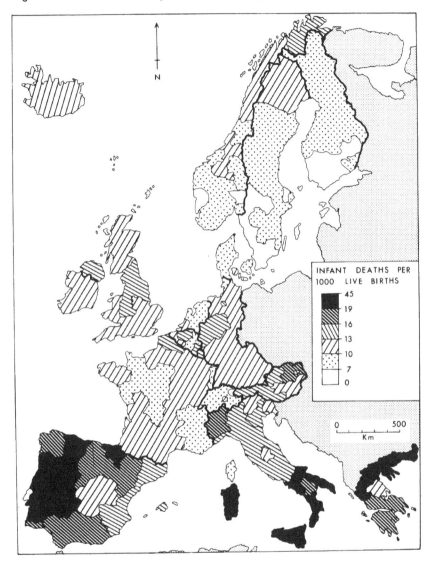

Source of data: National statistical abstracts.

Figure 2.2 shows the pattern of variation in 1980 across some 200 major provincial units. At this level of resolution there is a striking gradient between the best-off region (Värmland, Sweden: 3.9) and the worst-off (Trás-os-Montes/Alto Douro, Portugal: 43.1). In general, the lowest mortality rates are found in southern Scandinavia, with Akershus and Hordaland (Norway), Uppsala, Kronoberg, Gotland, Blekinge, Kristianstad, Halland, Gothenburg/Bohus, Älvsborg, Värmland and Örebro (Sweden), and Uudenmaan, Turun-Purin, Hameen and Mikkelin (Finland) all recording infant mortality-rates of less than 7.0. These must be very near to the 'irreducible minimum' level speculated about by community health specialists. Outside southern Scandinavia, only Appenzel (Switzerland) had a comparably low level of infant mortality, though much of the rest of Scandinavia, together with most of the Netherlands, southwest Belgium, northwest France, Switzerland and the Alpine regions of southeast France and northwest Italy had rates of less than 10.

At the other end of the spectrum are the disadvantaged regions of Greece, southern Italy and the Iberian peninsula. Although none of these approached the dreadful levels of mortality experienced in Trás-os-Montes/Alto Douro (which itself compares with the rate for Scandinavian countries in the 1920s), there were several regions where levels of mortality were more than double those for Western Europe as a whole. These include Minho, Beira Alta, Alto Alentejo and Douro Littoral (Portugal), León (Spain), Thráki (Greece) and Campania (Italy): 'peripheral' regions in 'peripheral' countries. Among the regions with intermediate levels of infant mortality, it is worth noting, there is a striking tendency for the major urban/industrial heartlands to stand out with higher rates than adjacent regions. The present pattern of infant mortality thus simultaneously exhibits a north-south dimension, a core-periphery dimension, and at least something of an urban-rural dimension. As we shall see, these are recurring themes in the social geography of Western Europe.

Fertility. The volatility of fertility-rates in Western Europe in the 'late' phase of the demographic transition has attracted a great deal of attention from demographers and planners because of the tremendous implications for demographic composition, economic development and public expenditure. In overall terms, the period since 1945 has been characterised by a steady increase in fertility followed by a marked decrease (Bourgeois-Pichat 1981). The peak year for birth-rates in many European countries was 1964. In that year there were over one million births in West Germany and less than 650,000 deaths; but by 1975 the number of births had fallen to less than 600,000 and the number of deaths had risen to 750,000. A natural increase of over 420,000 a year in the mid-1960s had been converted to a natural decrease of 150,000 a year in just over a decade. In Great Britain, over the same period, a natural increase of nearly 400,000 a year had changed to a small net decrease and similar, though numerically less dramatic, changes occurred in many other countries (Davis 1978). In the past few years there has been a slight upturn in fertility in some countries (Figure 2.3), though it is unlikely that it heralds as marked or as widespread a phenomenon as the rise in fertility in the early 1960s (Westoff 1983). The index of fertility used in Figure 2.3 is the total fertility-rate, which is not susceptible to variations in age structure, and which demographers regard as the most sensitive and meaningful cross-sectional measure of fertility (H.R. Jones 1981). The most common pattern revealed by this index is the so-called 'baby boom' immediately after the Second World War, when there was a catching-up of marriages and of births postponed during the war. Only Austria, West Germany and Switzerland did not experience this phenomenon, though it should be noted that the records for Ireland, Spain and Greece post-date this phase altogether. A second widespread pattern is the peak of fertility centred on the mid-1960s. This was most pronounced in Austria, England and Wales and Scotland, but it was common to all the countries shown on Figure 2.3 with the exception of Denmark, Finland, Greece, Portugal and Spain. The more recent upturn in fertility was first evident in England and Wales and Scotland in 1978, spreading to France, Sweden and Switzerland the following year and to Austria, the Netherlands and West Germany in 1980, and probably represents a bottoming-out of the fertility decline rather than a new, upward trend, as suggested by some (Ermisch 1982a).

These fluctuations are not simply the result of temporary changes in the timing of births and marriages. They are the product of complex changes in the entire pattern of family formation and reproductive behaviour in Western Europe; changes which are intimately related to many aspects of social and economic geography. The interrelations between the various mechanisms of social and demographic change are so complex and the variations in level and pattern from one country to another are so substantial that it is very difficult to identify a clear chain of events with distinct turning points. Nevertheless, the principal dimensions of change associated with postwar fluctuations in fertility have been documented by several writers (see, for example, Bourgeois-Pichat 1981; Coleman 1980; Council of Europe 1982; Monnier 1981). What follows is a broad outline of these changes in rough chronological order.

Beginning in the immediate postwar years, the average age at marriage fell significantly in every country except Ireland. Similarly, there began a sharp decline in the widespread spinsterhood which had been characteristic of Western Europe for the previous 200 years. By the early 1960s the average age at marriage was between 22 and 24, and it had become clear that fewer than 5 per cent of the women born around 1940 would remain unmarried for their entire lives (Festy 1980). Regional differences did persist, but they were mainly associated with ethnic sub-regions (where language barriers tend to circumscribe the marriage market) and with major metropolitan regions (where the continual influx of unmarried persons tends to depress nuptiality: see Watkins 1981). At the same time, the average interval between marriage and a first birth remained short, boosting the increased fertility associated with the postwar 'catching-up' of older cohorts of women, and leading to the sustained upward trend in fertility which is so pronounced in Figure 2.3.

The reversal of this trend in the mid-1960s is attributable to a number of factors. There appears to have been a widespread shift in preferences away from familism towards consumerism: a change which was undoubtedly fostered by the knowledge that reliable methods of birth control were available (Andorka 1978; Roussel and Festy 1979). The number of families with three or more children, for example, declined significantly throughout Western Europe between 1960 and 1980 (Table 2.1). Also, whatever the intentions of young couples at marriage, delayed parenthood inevitably fostered a taste for lifestyles based on the availability of two incomes; and as time passed the clear social and financial costs of starting a family began to outweigh, for some, the less tangible benefits of having children. What started out as deferred births thus became, in some cases, cancelled births. At the same time, changing social attitudes about the status of women came to be reflected in improved educational opportunities and a wider choice of employment, both of which fostered the development of non-familistic lifestyles. The consequent decline in birth-rates was reinforced soon afterwards in many countries by changes in abortion legislation which made it easier to terminate unwanted pregnancies. In addition, once the proposition that sex need not be aimed primarily or solely at procreation had become generally accepted, further trends were set in motion. The social value of marriage decreased, with a consequent decline in the rate of marriage, an increase in the average age at marriage, an increase in divorce, an increase in cohabitation without marriage, and a decrease in the rate of illegitimate births – all conspiring to depress the fertility-rate still further (Calot and Blayo 1982; van de Kaa 1980). These trends appeared first in Denmark and

Sweden around 1965 and have since spread progressively to Switzerland, West Germany, Great Britain, Norway, France and Italy.

Figure 2.3: Fertility Trends in Western Europe, 1945-80

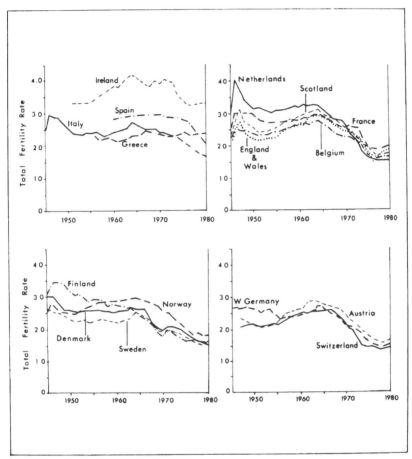

Source of data: National statistical abstracts.

The critical question for the immediate future is whether this spiral of downward fertility will be checked. As we have seen, several countries have recently experienced the beginnings, at least, of an upturn in fertility. The limited evidence which is available at present, however, suggests that this is mainly the result of a rise in the fertility-rates of women in their late twenties and early thirties (Muñoz-Perez 1982), and it must be doubtful whether the supply of these deferred births will be sufficient to counterbalance the decreased fertility resulting from postwar changes in the overall pattern of family formation. Meanwhile, every country except Greece, Portugal and Spain had a total fertility-rate in 1980 which was less than 2.1, the theoretical level of 'replacement fertility'. This has renewed speculation about the existence of a fifth stage in the demographic transition: a senile stage with fertility falling consistently below the death-rate, causing population decline.

Table 2.1: Third and Later Births as a Percentage of Total Births, 1960-80

	1960	1980
West Germany	28.2	17.1
France	38.8	21.2
Italy	34.8	20.2
Netherlands	41.8	19.8
Belgium	37.3	18.6
Luxembourg	27.8	15.9
UK	33.2	22.3
Ireland	60.9	46.2
Denmark	36.0	17.1
Greece	27.4	17.7
Portugal	45.2	23.9
Spain	—	30.3

Source: Eurostat (1982).

Some Implications of Slow Population Growth

This concern about population decline is by no means a new pheno-menon, though the perceived implications of population decline have changed a good deal. In the latter part of the nineteenth century, when decreasing birth-rates were accompanied by the development and dissemination of genetic theories which linked social behaviour to heredity, concern was focused on the differential fertility which has always existed between social classes in northwestern Europe. Popula-tion decline, it was argued, would lead to an erosion of the inherent *quality* of the population as the educated elite failed to reproduce itself in sufficient numbers. Worse still, for some, was the prospect of an even greater numerical dominance by the more fertile proletariat, with all the associated implications for political change (Winter 1980). During the early part of this century population decline came to be viewed by some as a form of unilateral disarmament, a feeling which was heightened by the political climate of the interwar period. Not surprisingly, many governments introduced pro-natalist policies, the most explicit of which were those pursued in Nazi Germany and Fascist Italy. These involved the suppression of information about contraception, criminal prosecution for induced abortion, a tax on single persons, marriage loans (which could be partly written off by having children), family allowances and tax concessions for large families (de Sandre 1978; Glass 1967). Less draconian policies were pursued in Belgium, Sweden and the United Kingdom, where concern about population decline was couched more in terms of the downward spiral of investment, consumption and living standards which was expected to follow population decline (see, for example, Reddaway 1939).

It has been France, however, which has nurtured the darkest and most obsessive fears about population decline. As many writers have pointed out, this is largely the result of its exceptional demographic history. An extremely early start to fertility decline meant that while Britain's population grew by 43 per cent between 1871 and 1911 and that of the German Empire grew by 58 per cent, France could manage less than a 10 per cent increase (Dyer 1978). Following the loss of over 1.4 million people during the First World War, the French were intimidated by the rising birth-rates in Nazi Germany and Fascist Italy, and dismayed when the death-rate began to exceed the birth-rate in 1935. Already there had developed a wide spectrum of pro-natalist groups, and in 1939 the *Code de la Famille* was instituted, reinforcing restrictive measures against the use of contraception, extending family allowances, introducing tax breaks encouraging family formation, and imposing restrictive legislation relating to abortion. After the Second World War, as we have seen, there was in any case a general increase in fertility as a result first of the 'baby boom' and then of changing patterns of nuptiality; but it has been suggested that the vigorous pursuit of pro-natalist policies and propaganda in France may itself have raised fertility by as much as 10 per cent (Calot and Hecht 1978). Nevertheless, the French continue to frighten themselves with statistics, constructing elaborate projections of the consequences for the population if a particular level of fertility were to be maintained: 'As soon as fertility falls below replacement level, there is a rush to calculate the day on which the last Frenchman will be born' (Ogden and Huss 1982: 284). Not surprisingly, French demographers have been prominent in the debate surrounding the possible onset of a new 'senile' stage in the demography of Western Europe. But, while the French may have retained some of their fears about the impact of slow growth on the role of their culture and political influence in the world (see, for example, *Après-Demain* 1980), the real focus of concern has shifted towards issues concerning economic development, the maintenance of welfare systems, and the changing pattern of household composition (Council of Europe 1978; United Nations 1975).

Population Stagnation and Economic Development. The first thing to be said about the relationship between population change and economic development is that it is imperfectly understood (Chesnais and Sauvy 1973; Ohlin 1967). It should also be acknowledged that population decline may, in the long term, be a blessing in disguise, bringing a better balance between population and resources. Nevertheless, the slowing-down of population growth does have clear implications for the future pattern of consumer demand in the short and medium term and, therefore, for national and regional economic well-being. Maillat (1978) has discussed these issues, emphasising the point that it is not so much the total number of consumers which is important as the changing structure of needs and demands arising from changes in demographic structure. One important example is the fall in demand in the building and construction industries and the reduction in the need for certain infrastructural components (schools, for example) associated with a fall in household formation.

Perhaps the most widely recognised of the economic aspects of population decline, however, are those related to the supply of labour (Ermisch 1982b). The rate at which the labour force in any one country is able to renew itself is a particularly crucial aspect of economic health (Ryder 1975). Although in recent years the increased supply of female labour, the availability of migrant labour and the effects of the international economic recession have more than made up for any shortages of labour, Guilmot (1978) has shown that the underlying trends point to many Western European countries eventually being unable to replace their labour force. Because of the general rise in fertility between 1945 and 1964, however, there will be no immediate problem of replacement. (Indeed, the product of the peak birth-rate in the mid-1960s is only now entering the labour market, representing a 'disadvantaged cohort' of European population which seems likely to find competition that much greater not only in the labour market but also in housing markets and many other spheres of life.) There are some aspects of the changing nature of labour supply which are more pressing, however. These are explicitly related to the shifting age structure of the labour force. While it is very difficult to weigh the occupational experience and reliability of the 'old' against the dynamism and innovation of the 'young' – especially when technological change and structural economic

trends are taken into account − it is generally accepted that the productivity of both young (an increasing component of the workforce in the
recent past) and old (an increasing component of the workforce in the
near future) is significantly lower than that of 20 to 50-year-olds
(Guilmot 1978). Another immediate problem arising from the changing
age structure of the workforce centres on the issue of career structures.
As Keyfitz (1973) has shown, the workforce in slow-growing or stable
populations tends to suffer from an acute restriction of opportunities
for promotion at intermediate levels, thus increasing the number of
cases in which careers reach an 'early peak', followed by a long
stationary or declining period. This, in turn, is likely to have important
implications in terms of consumer demand, lifestyles, and attitudes to
work.

Population Stagnation and Social Welfare. The corollary of changes in
the size and composition of the labour force may be seen in the changing size and composition of the dependent population: schoolchildren,
housewives without paid employment and retired persons. There has
been for some time now a widespread trend within Western Europe
towards prolonged education and training and earlier retirement, thus
tending to expand the potential size of the dependent population (Kirk
1981). Figure 2.4 shows the regional pattern of dependency, as measured
by the simple ratio of the economically active population to the inactive
(i.e. dependent) population. Throughout most of Western Europe there
is at least one dependant for each economically active (but not necessarily employed) person; but in southern Italy, Sicily, Sardinia, south-
western Spain, eastern Austria and the northern Netherlands the
dependency ratio is twice as high, emphasising the general north-south
gradient of the map.

    Because of postwar trends in fertility, the dependency ratio began to
decrease after 1970 and will continue to do so at least until the year
2000 (Guilmot 1978). Nevertheless, present levels of dependency represent a considerable burden on the economically active population and
have proved to be a considerable strain on the state welfare systems
which have been established in many countries since 1945 (Wirz 1977).
Moreover, much of the anticipated decrease in dependency ratios will
result from a reduction in the proportion of children, who cost only
half as much as the elderly in terms of public expenditure (Bourgeois-
Pichat 1981). It follows that the decrease in dependency ratios will not
be matched by proportional decreases in the burden they impose. On
the contrary, the dependent population will almost inevitably exert an

increasing burden as it becomes increasingly concentrated in the older age groups which are more expensive to maintain (Figure 2.5). This is something which has attracted a good deal of attention among social scientists and policymakers (Abrams 1979; Eversley 1978, 1982; Feichtinger 1975; McDonald and McDonald 1982; Stearns 1977). One of the major concerns is the question 'Who will pay our pensions?' (Chadelet 1975). This, of course, is an issue which mainly concerns national welfare systems and large-scale private pension schemes. Many of the implications of ageing populations are set at the community level, however, contributing an important dimension to the social geography of contemporary Western Europe. Ageing populations impose increasing demands on a wide spectrum of local authority services. One example is the provision of old people's homes. In Denmark, for instance, 16,000 extra places will be required in old people's homes by the year 2000 if the levels of provision achieved in 1980 are to be maintained. Should mortality among the elderly continue to decline at the overall postwar rate (and there is no reason to suppose it should not), the need will be for 27,000 extra places (Leeson 1981). Moreover, these figures emphasise that the issue is not simply based on quantitative changes. There is also a qualitative aspect to change, involving an increased number and proportion of the geriatric elderly. Other aspects of public service provision involved in the 'burden' of ageing populations include specialised health facilities and health care delivery systems, specialised housing and domiciliary services, transportation and environmental design, and neighbourhood and community care (Amman 1981; Bergman 1979; Miflas 1980). All involve increased levels of expenditure; they also imply changes in social priorities and changes in the nature and organisation of service delivery. This, in turn, adds an important political dimension to the whole issue.

Population Stagnation and Household Composition. While it must be accepted that changing patterns of household composition are a product of a variety of factors it is clear that demographic trends exert a strong influence on patterns of family status (see, for example, Haavo-Manila and Kari 1979). Two particular consequences of this relationship are relevant here, since they are both related to population subgroups which are not only socially marginal and economically vulnerable, but which have also become localised in specific regions and neighbourhoods.

Figure 2.4: Dependency in Western Europe *c.* 1980

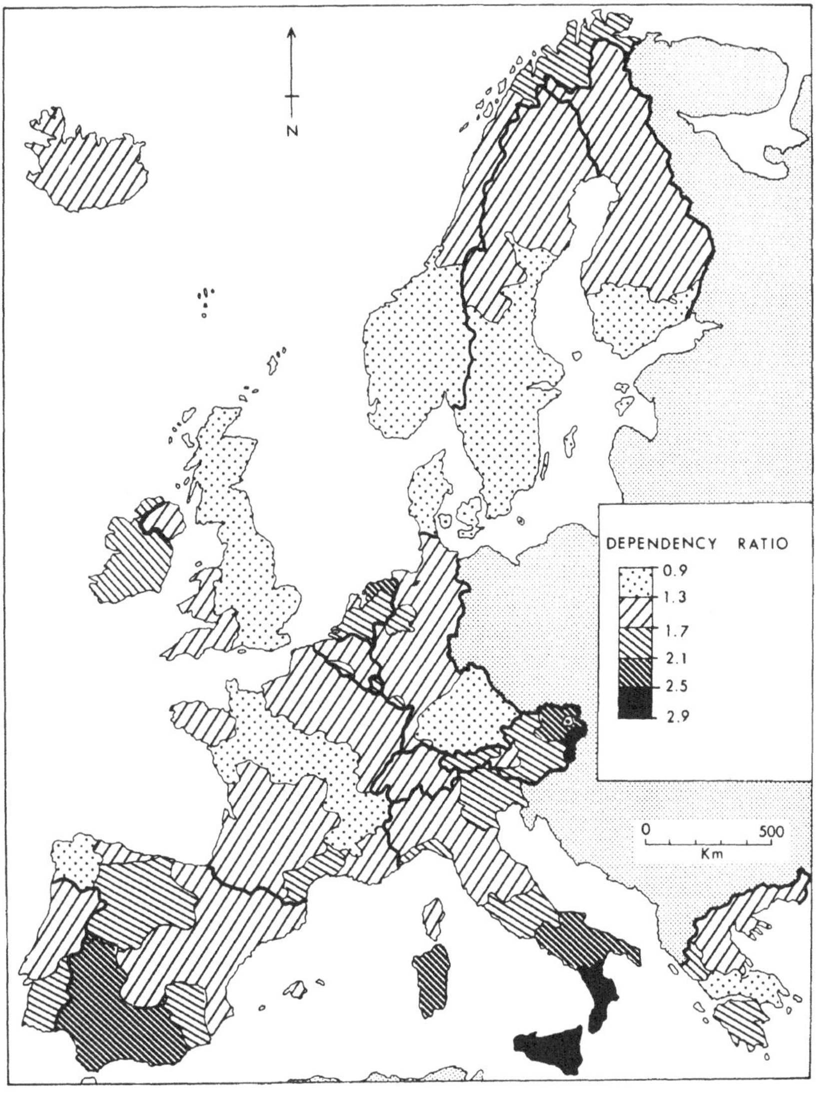

Source of data: National statistical abstracts.

The first of these sub-groups involves the lone elderly. Because of the differential between male and female mortality (a gap of six years in terms of life expectancy at birth), the ageing profiles of stationary or declining populations produce increasing numbers of widows. Thus, for example, if 8 per cent of married women are widows in a population growing at a rate of 2.8 per cent per annum, the figure will be 25 per cent in a stationary population having the same rate of nuptiality and mortality (Le Bras 1974). With the postwar trend away from familistic lifestyles and the consequent attenuation of kinship systems, the lone elderly have become increasingly isolated, both personally and socially. Moreover, they are becoming increasingly localised in specific communities (inner-city neighbourhoods, interwar housing developments and small villages, for example) as a result of the dynamics of residential segregation (see Chapter 6) and regional migration (see below, pp. 36-7).

Figure 2.5: Changing Age Composition of the European Population, 1800-2050 AD

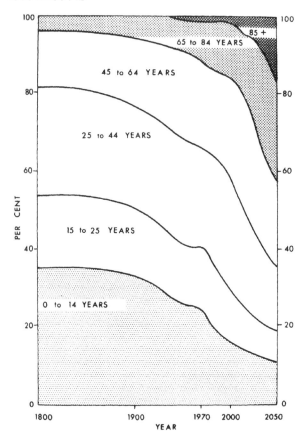

Source: Bourgeois-Pichat (1981).

The second sub-group involves the single-parent family. As Clason (1980) points out, the increasing importance of the single-parent family must be seen as a product of ideological, economic and technological changes as well as the outcome of postwar demographic trends. The most important factor, though, is the increasing rate of divorce associated with the post-1964 demographic downturn. Over the last ten years or so, the incidence of divorce has doubled or even tripled in many parts of Western Europe (Festy 1980). Admittedly, a fair proportion of divorcees with children eventually remarry. Nevertheless, the increased rate of divorce has produced much greater numbers of single-parent families, especially in northwestern Europe. Together with lone unmarried mothers, these families represent a particularly vulnerable group: in Britain, for example, 45 per cent of all single mothers have to rely on the welfare state's income-maintenance programme (Pierce 1980). Like the single elderly, they also tend to become localised, particularly in inner-city environments, contributing to distinctive syndromes of stress and deprivation.

Migration and Social Well-being

The interplay between migration and economic development has left its imprint in a variety of ways across the whole of Western Europe. Indeed, the various streams and currents of migration which have taken place since the onset of the urban/industrial transition have been a major component in the dynamics of regional development, interacting with social, cultural and political change as well as with economic change. Traditionally, this component had been interpreted as part of an equalising mechanism, moving labour from surplus to deficient regions as workers migrate in search of higher incomes. As a result of such movements, together with movements of capital and enterprise, wage differentials eventually narrow, leading in turn to a convergence of regional and national well-being. This is the basis of neoclassical equilibrium theory (see pp. 197-8). Such a view of migration is now regarded, at best, as naïve; at worst, as misleading. Post-neoclassical interpretations have for some time recognised that real people do not necessarily behave rationally, that there is a tendency for political, bureaucratic and monopolistic factors to constrain 'free-market' behaviour, and that it is regional *rates of growth* rather than differences in the absolute level of incomes which best explain migration streams (Gober-Meyers 1978). To a large extent, however, the conventional wisdom concerning migration and economic development has shifted to the very different theoretical framework offered by Myrdal (1957) and Hirschmann (1958), who emphasise that migration often serves to maintain, or even exaggerate, the economic and social disparities of which it is initially a function. As such, migration is seen as one of the key mechanisms of 'cumulative causation', whereby the initial advantage held by prosperous nations and regions is cumulatively reinforced by flows of capital, enterprise and skilled labour. At the same time, the loss of these basic factors of production serves to launch disadvantaged nations and regions on a downward spiral. More recently, the relationship between migration and economic development has been further reappraised, this time using Marxist economic theory as the basic framework for interpretation. According to this view, migration is seen as one of a whole spectrum of mechanisms whose net effect is to subordinate disadvantaged or 'peripheral' regions to 'core' regions (see p. 198). In this context, migration reflects the mobilisation of an 'industrial reserve army' of labour in dependent, peripheral regions in response to the changing imperatives of capital accumulation (Gustaffsson 1979).

All three views clearly emphasise, in one way or another, migration as a response to the dynamism of socio-economic development. Nevertheless, all of them also recognise that migration is itself a catalyst and a stimulus for social and economic change. In addition to changes in the well-being of migrants themselves, there are two principal aspects of this change: one involving the welfare of recipient communities, the other involving the welfare of sending communities. The *nature* of such change is multidimensional, however. The impact of emigration on sending communities, for example, involves both positive and negative consequences of several different kinds. One positive aspect of out-migration which has been widely cited is the short to medium-term palliative effect of siphoning off unemployed and underemployed labour. Another is its stabilising effect in channelling out the frustrated and so defusing political tension. On the other hand, it has been argued that even the unemployed maintain a certain level of expenditure, so that their outmigration results in a decline in local purchasing power and, ultimately, a further increase in unemployment (Vanderkamp 1970). More important, perhaps, is that outmigration has been found to be highly selective not only in siphoning off the economically and demographically active population, but also in removing the better-qualified. Similarly, the economic-multiplier effect of inmigration has to be balanced against the social tensions and public welfare needs generated by immigrant communities. The following sections describe the major dimensions of European migration before going on to examine some of these welfare-related issues in more detail.

Major Streams of Migration Between 1850 and 1950

The disruptions and dislocations associated with the transition from a rural to an urban/industrial society, together with the upheavals of two world wars and a number of other political crises, brought about unprecedented movements of population within Western Europe between 1850 and 1950. These movements were crucial not only in restructuring the population geography of the region, but also in shaping its economic landscape and modifying its political and cultural complexion. The flows and counter-flows involved were complex, but it is useful to recognise four major categories:

(1) *Intercontinental migration* Partly in response to population pressure arising from the early phases of the demographic transition, and partly in response to the poverty and squalor of the early phases of industrial urbanism, vast numbers of people left Europe for overseas destinations. Beginning in the British Isles in the early nineteenth century, this emigration roughly paralleled the diffusion of the demographic transition and the industrial revolution, affecting central and northwestern Europe by the late nineteenth century and southern Europe by the early twentieth century. At the peak, towards the turn of the century, emigration from Europe (including Eastern Europe) amounted to about a million persons each year; in total, some 55 million to 60 million people left Europe between 1850 and 1950, over one quarter of them from the British Isles (Kosiński 1970). The main stream of migration was to the Americas, with people from northwestern and central Europe heading for North America and southern Europeans heading for destinations throughout the Americas. In addition, large numbers of British left for Australasia, eastern and southern Africa, while French and Italian emigrants made for North Africa, Ethiopia and Eritrea, and the Dutch made for Indonesia. The final surge of emigration occurred just after the Second World War, when various relief agencies helped homeless and displaced persons to move to Australasia, North America and South Africa, and large numbers of Jews relocated in Israel.

(2) *Movement to the coalfields* The main change in patterns of population distribution *within* Western Europe came with the shift in economic organisation from the rural/craft base to the industrial base. This shift required a factory system of production operating with large, specialised labour forces; it also drew heavily on coal-based energy systems. Thus,

although the onset of industrial capitalism brought with it a major expansion of existing urban systems (and particular growth in ports and capital cities), the *regional redistribution* of population was largely determined by the location of coalfields. The areas affected most by this redistribution were of course those in northwestern Europe, which were first to experience the industrial revolution. In Britain, southeast Lancashire and west Yorkshire had by 1850 already been transformed from poorly settled regions to nascent conurbation centres. Soon afterwards these were joined by south Yorkshire, northeast England, South Wales, Clydeside and the West Midlands, together with the Ruhr area and northeast France/southwest Belgium. Meanwhile, it was the less prosperous rural areas, together with the obsolescent small market towns and guild towns of the old economic order, which were the chief sources of supply of migrants. The legacy of coalfield-oriented migration can thus be seen not only in terms of the urban/industrial landscape, but also in terms of the many picturesque small towns which escaped the direct effects of industrialisation.

(3) *Drift to metropolitan cores* The dominance of the coalfields ebbed with the development of new sources of energy, new technologies and new markets (see Chapter 5). As industrial capitalism changed and adapted, it was the diversified economies of national metropolitan centres which offered most opportunities and the highest wages, thus prompting a further redistribution of population. In Britain this involved a 'drift south' to London and the Home Counties; in France it resulted in a polarisation between Paris and the rest of the country. Some nations, drawn into the orbit of industrial capitalism after the 'coalfield' stage, experienced a more straightforward shift from peripheral rural areas to prosperous metropolitan core regions. Thus emerged the intensification of metropolitan dominance around Barcelona and Madrid, Milan and Turin, Oslo, Copenhagen and Stockholm. Meanwhile, the whole process was reinforced by a continuation of the general rural-urban drift which had been implicit in the coalfield-oriented movement.

(4) *Forced migrations* Overlying the three streams of intercontinental, coalfield and metropolitan migration were forced migrations associated with wars and political crises. According to Kosiński (1970) the First World War forced about 7.7 million people to move, though the bulk of this activity took place in Eastern Europe. Nevertheless, the reclamation of Alsace and Lorraine (by France) and Eupen-Malmédy (by Belgium) prompted several thousands of families to return to Germany, where thousands more had returned from Poland and Czechoslovakia.

Similarly, the disintegration of the Austro-Hungarian Empire prompted nearly 800,000 people to withdraw to Austria from prewar locations in Czechoslovakia, Germany, Hungary, Italy, Poland and Yugoslavia. Another major transfer of population took place in the early 1920s, when over one million Greeks were repatriated from Turkey in the aftermath of an unsuccessful Greek attempt to gain control over the eastern coast of the Agean Sea; 300,000 Turks, meanwhile, moved in the opposite direction. Soon afterwards, more people were on the move in the cause of ethnic and ideological 'purity' as the policies of Nazi Germany and Fascist Italy began to bite. Jews, in particular, were edged out of Germany, Italy and Spain. Nearly 90 per cent of the 400,000 refugees from the 'German Lands' between 1933 and 1939 were Jewish. Many emigrated to Palestine or North America, though a substantial minority settled in France, Britain and other Western European countries. Ironically, the last of the forced movements of population before the Second World War involved 72,000 German-speaking South Tirolers who opted for a return to Germany rather than accept Mussolini's alternative of 'compulsory assimilation' in Italy. With the Second World War there occurred further forced migrations on a huge scale (25 million people were involved this time), though again the bulk of the movement took place in Eastern Europe. During the war itself, the *Volksdeutsche* (ethnic Germans) were gathered back to the fatherland from the Baltic states and east-central Europe. After the war the political cleavage between Eastern and Western Europe resulted in major transfers of population, many of them focused on Germany. By 1950 West Germany had absorbed nearly 8 million refugees, most of them ethnic Germans from Czechoslovakia, Hungary and Poland. These were joined by a further 3 million refugees from East Germany before the Berlin Wall was built in 1961.

Recent Currents of Migration within Western Europe

In addition to the refugee movement associated with the aftermath of the Second World War, most countries experienced a flurry of regional migration in the immediate postwar years. Subsequently, it became clear that new patterns of migration were beginning to emerge (Thumerelle 1981). The drift towards metropolitan core regions has continued to be a strong component of postwar migration (Kayser 1977), as has rural-urban migration (Clout 1976; Romanos 1979); indeed, it should be emphasised that 'gravitational' factors have been the dominant influence on regional migration flows. In addition, intercontinental emigration has continued as a significant influence on the population geography of Western Europe, though from 1950 it became a more selective 'brain drain' of technicians, scientists, teachers and medical personnel to North America and Australasia.

Other streams of migration have joined these in regrouping Europe's population, however:

(1) *Decentralisation* The trend towards suburbanisation and metropolitan decentralisation had already begun during the interwar years, but since 1950 it has grown to become a major feature of the social geography of Western Europe. Essentially, it is a reflection of the changing structure of urban systems, and it is discussed in detail in this context in Chapter 6.

(2) *Retirement migration* Again, this is a phenomenon which is by no means new. Since 1950, however, it has gained considerable momentum. Its importance here lies in the localisation of its impact in particular regions. Two trends seem to be operative. One involves a return to the place or district of origin of retired persons who had previously migrated in pursuit of economic opportunities. The other involves migration, on retirement, to 'places of reward and repose': usually places that have been scouted during vacations some years before (see, for example, Karn 1977; Law and Warnes 1981). Both trends result in a clear urban-rural component, though it is the second type of movement which carries the greatest spatial impact, creating concentrations of the elderly in resort-cum-retirement regions such as Cornwall and the Lake District in England. Such regions, together with the spas and coastal resorts which have also attracted the elderly, represent a distinctive dimension of European social geography which has received surprisingly little attention from geographers. It is clear enough from the existing evidence, however, that while most of these regions, resorts and spas exhibit some common attributes, these exists a wide spectrum of 'retirement communities' (Law and Warnes 1980).

It is also clear that retirement migration is complex in its selectivity. In a study of French retirement migration, for example, Cribier (1974) found that there was in general a strong correlation between education and propensity to migrate on retirement. At the same time, clear differences emerged according to place of residence: one in three of those in higher-status groups in Paris tended to stay on, whereas three out of every four lower-status Parisians left, mainly because of the difficulties of living in such a high-cost environment on modest pensions. In contrast, the higher-status elderly in provincial cities were more likely than the elderly working classes to migrate: partly because they had the means to pursue 'wider geographical horizons'. Cribier's (1974) work also demonstrates the selectivity of retirement migration in terms of destination:

> Well-off Parisians are three times more likely than the lower class to retire in an area from which they do not originate, and settle more often in the 'second homes belt' around Paris, in resort towns of the Atlantic and Mediterranean coasts and in the foothills of the Mediterranean mountains. They are particularly numerous also in the favoured regions of the middle Loire, in the Basque country and more recently in Perigord. (p. 372)

Even when they return to their area of birth, better-off Parisians behave very selectively. Only 10 per cent of those originating in northeast France return there, compared with 50 per cent of those originating in the Midi.

(3) *Migrant labour* Although estimates vary a good deal, there were at the beginning of the 1980s some 14 million aliens living in Western European countries. About half of these were young adult males who had emigrated in order to seek work, originally, at least, on a temporary basis. The rest were wives and families who had joined them. At the heart of this inflow of foreign-born workers lie the labour needs of the more developed countries, coupled with the demographic 'echo effect' of low birth-rates in the 1930s and 1940s — a sluggish rate of growth in the indigenous labour force of northwestern Europe (McDonald 1969; Drewer 1974). As the demand for labour in more developed countries expanded, so indigenous workers found themselves able to shun low-wage, unpleasant and menial occupations (Böhning 1972; Castles and Kosack 1973). These jobs represented welcome opportunities, however, to many of the unemployed and underemployed workers of Mediterranean Europe and the more distant realms of former European colonies. At the same time, the more prosperous countries perceived that foreign workers might provide a buffer for the indigenous labour force against the effects of economic cycles — the so-called *konjunkturpuffer* philosophy.

Britain and France were the main recipients of immigrant labour from the Third World. Commonwealth citizens began to arrive in Britain at the rate of about 30,000 per annum in the mid-1950s, first from the Caribbean and then from India and Pakistan. Once the slack in the labour market had been taken up, however, there emerged a public reaction against immigration, with the result that the 1962

Commonwealth Immigrants Act introduced a voucher system which closely regulated the number of new workers admitted while allowing for the entry of dependants of workers already in Britain. Between 1955 and 1968 some 669,640 Commonwealth immigrants were admitted to Britain (Deakin 1970). In France, the Office National d'Immigration was set up immediately after the war to organise and control the entry of immigrant workers. Algeria provided the bulk of non-European recruits at first, amounting to over 250,000 by 1954 (Trébous 1980). In the next fifteen years, despite immigration restrictions introduced in 1964, the number of Algerians in France increased to over 600,000. Meanwhile, streams of immigrants also began to arrive from elsewhere in North Africa and from Senegal, Mali and Mauritania in West Africa (Castles and Kosack 1973).

Table 2.2: Foreign Workers (in thousands), Selected Countries, 1975-6

Emigrant countries	Immigrant countries					
	Belgium	West Germany	France	Sweden	Switzerland	UK
Austria	1.0	76.0	–	3.0	20.4	–
Greece	10.0	178.8	–	8.5	–	2.5
Italy	96.0	276.4	199.2	2.9	261.6	56.5
Portugal	6.0	69.6	360.7	1.0	–	4.0
Spain	30.0	111.0	204.0	1.8	67.9	15.5
Turkey	16.0	527.5	31.2	3.5	–	1.5
North Africa	33.5	17.0	556.4	1.0	–	1.5
Yugoslavia	3.0	390.1	42.2	26.1	24.1	3.5
Others	121.3	290.7	190.6	177.5	142.0	690.0[a]
Total	316.8	1937.1	1584.3	225.3	516.0	775.0

Note: a. 'Others' for the United Kingdom includes 480,000 born in Ireland.
Source: Kirk (1981).

Britain and France, together with the rest of the more prosperous nations of northwestern Europe, also received substantial numbers of immigrant workers from less developed countries within Western Europe (Table 2.2). Italy was one of the earliest major suppliers of migrant labour, soon to be followed by Greece, Portugal, Spain, Turkey and Yugoslavia. The peak of these streams of migration occurred in the mid-1960s, with a secondary peak at the beginning of the 1970s, reflecting the peaks of economic cycles (Calvaruso 1982). The onset of the deep general recession in 1973, however, brought a dramatic check to the flow of migrant labour (OECD 1975). Restrictions on the admission of non-EEC immigrants began in West Germany in November 1973, and within twelve months France, Belgium and the Netherlands had followed with new restrictions. By this time there were nearly 2 million foreign workers in West Germany, over 1.5 million in France, three-quarters of a million in the United Kingdom, half a million in Switzerland, and around a quarter of a million in Belgium and Sweden (Table 2.2). These numbers can be roughly doubled if dependants are included, though precise figures are difficult to obtain because of the unknown quantity of illegal immigration. Within receiving countries the demographic impact of migrant labour has been localised in the larger urban areas, reflecting the immigrants' role as replacement labour for the low-paid, assembly-line and service-sector jobs vacated in inner-city areas by the upward social mobility and outward geographical mobility of the indigenous population (Salt 1976). Thus, for example, more than one-third of the 4.1 million aliens in France in 1975 lived in the Paris region, representing 14.5 per cent of its population. Other French cities where aliens account for more than 10 per cent of the population include Lyon, Marseille, Nice and St Etienne (Kayser 1977). In West Germany the increasing concentration of foreign workers in large cities (except Berlin) led to the imposition in April 1975 of quotas which limit the numbers of aliens living in city centres (Kirk 1981).

Within the sending countries, on the other hand, the demographic impact of these migration streams has generally intensified the flow of rural outmigration from the least developed regions: southern Italy, the islands and mountainous northern region of Greece, northern Portugal, and southwestern and northwestern Spain. Galicia (in northwestern Spain) and Andalusia (in the southwest), for example, have supplied 55 per cent of all Spanish migrants to Europe since 1960 (Kirk 1981). It should also be recognised that much of this outmigration has involved complex but very specific streams of movement from one region to another. Thus, for example, although in 1970 France received less than 10 per cent of all Italian emigrants to Europe, almost a third of those leaving Liguria and over a quarter from Piemonte went to France. Meanwhile, Sicilia, Sardinia and Veneto sent over half their emigrants to West Germany, and Lombardia sent nearly 80 per cent of its emigrants to Switzerland (Salt and Clout 1976).

(4) *Return migration*. After the onset of economic recession in 1973, return migration became a major feature of European migration streams. In France, a donation of £1,100 and a free ticket home was offered as an inducement to repatriation (Kennedy-Brenner 1979), and 650,000 foreign workers left West Germany in 1975 as the *konjunkturpuffer* approach took effect. As a result, 'exporting' countries soon found themselves as net importers of labour. Table 2.3 shows the dramatic

turnaround in the flows between West Germany and its major suppliers of *Gastarbeiter* ('guestworkers') between 1973 and 1974. By 1974 all of the 'suppliers' had become net importers of labour, despite continued outflows of migrants in response to persistent problems of rural unemployment.

Table 2.3: Flows of Non-German Migrant Workers into and out of West Germany, 1973 and 1974

	1973			1974		
	In	Out	Balance	In	Out	Balance
Greece	14,309	27,014	−12,705	6,838	26,375	−19,537
Italy	109,530	90,931	+18,559	53,185	82,889	−29,704
Morocco	3,404	994	+2,410	698	1,103	−405
Portugal	29,739	7,935	+21,804	2,376	12,991	−10,615
Spain	33,170	28,586	+4,584	4,223	33,425	−29,202
Tunisia	3,014	1,075	+1,939	675	1,397	−722
Turkey	131,437	40,508	+90,929	28,967	51,281	−22,314
Yugoslavia	108,368	67,513	+40,855	32,983	69,809	−36,826
Total	432,971	264,556	+168,145	129,945	279,270	−149,325

Source: Entzinger (1978).

Return migration is not simply a product of adverse economic conditions, however. Most returnees cite non-economic factors such as discrimination, homesickness, family ties, children's education, and the desire for enhanced status on return (Gmelch 1980). Moreover, return migration is a long-standing phenomenon which Ravenstein (1885) recognised in his Fourth Law of Migration a hundred years ago: 'Each main current of migration produces a compensating counter-current' (p. 33). Most interregional migration streams within European countries have thus generated a counter-stream. Taylor's (1979) study of migrants from West Durham who moved to the English Midlands during the 1960s, for instance, found that nearly 20 per cent of the families returned within a few years. Similarly, intercontinental emigration from Europe has generated a series of reverse flows. Among those which have been documented are those from Australasia (Price 1963), Canada (Beyer 1961; Richmond 1968) and the United States (Backer 1966; Handlin 1956; Saloutos 1956; Vagts 1960). Such flows have given rise to a number of critical economic and social problems which are discussed in more detail below (pp. 45-7).

(5) *'Rural turnaround'*. In recent years evidence has emerged which suggests that parts of Europe may be experiencing the kind of

turnaround which Beale (1977) claimed to recognise in North America. One important feature of the so-called rural turnaround is that some of the most spectacular rates of inmigration have occurred in areas which are relatively remote from metropolitan influences. In Britain, for example, peripheral rural areas in Wales and in the Borders and Highlands and Islands of Scotland increased in population by about 10 per cent between 1971 and 1981 whereas they had been losing population over the previous decade. The more prosperous rural areas of southern and eastern England, meanwhile, experienced a sharp decrease in their rate of growth (Champion 1981, 1982). It is also important to note that the migration streams which have caused these turnarounds in rural population trends seem to involve two very different groups: returnees and incomers. The significance of this distinction lies in the very different aspirations, orientations and skills of the two groups. There are therefore important differences between the socio-economic impact of returnees and that of incomers; and it is in many ways the tensions between the indigenous population, returnees and incomers which are the key to the contemporary social geography of many rural districts (see pp. 103-10).

Migration, Marginality and Social Conflict

The complexity of the interactions between migration, economic development and the well-being of migrants has already been stressed. In this section, some of the socio-geographic consequences of recent patterns of migration are illustrated in more detail. Three issues have been selected for this purpose: the socio-economic marginality of small communities affected by heavy outmigration, the problems associated with migrant labour in host communities, and the impact of the migrant-labour system on source areas.

Marginality. The economic and social viability of sparsely populated rural areas, small market towns and specialist fishing, mining and industrial villages is seriously weakened by outmigration. Such communities are 'marginal' in that they find themselves close to the critical threshold population necessary to support even basic amenities and services; they are also marginal in the sense that kinship and social networks become difficult to maintain, thus threatening 'community' with social instability. It will be recognised by now, of course, that outmigration is not necessarily synonymous with depopulation. Conversely, as Clout (1976) points out, depopulation may result from any of several circumstances: an excess of deaths over births, an excess of outmigration over immigration, migratory loss exceeding natural increase, or net outmigration combining with natural decrease. Nevertheless, it is clear that outmigration will be a key determinant of marginality wherever birth-rates are moderate or low, as in most of the more developed countries of northwestern and central Europe.

The *selectivity* of outmigration is a major factor in increasing the vulnerability of residual communities. Because of the age and sex selectivity of outmigration, it has the effect of reducing rates of natural increase. This, in turn, reinforces the trend towards depopulation. It has also been suggested that outmigration is selective in siphoning off potential innovators (Jones 1965) and social leaders (Capo and Fonti 1965; Dell'Atti 1979). From this point onwards, a self-perpetuating downward spiral of events is set in motion, the ultimate result of which may be to topple the community over the threshold of survival. Outmigration creates a vacuum in the local housing market which not only prompts a chain of residential mobility amongst local families, but also opens the way for a few outsiders: retired persons, perhaps, or pioneer ex-urbanites. As Taylor (1979: 485) puts it, 'The frequent presence of removal vans, the departure of friends and "known faces" and the appearance of "strangers", often fosters a kind of introverted conservatism in the ageing, residual community. The communality of community life is progressively eroded and is replaced with an air of despondency. As population dwindles it also becomes increasingly difficult to sustain shops and amenities; the quality of existing shops and personal services tends to decrease as demand falls away; and the reduced tax base makes it difficult to maintain even the most elementary public services (White 1980). The whole problem is compounded by the fact that the elderly tend to spend less in any case, especially on food. At the same time, they require more in terms of specialised public services. In short, such communities become less and less attractive places to live in. For many families the final stimulus to leave comes with the closure of the local school – a loss which is commonly thought to signal the final stages of community life. In fact, public subsidies and regional policies from central government have probably saved many marginal communities from reaching this critical threshold. Nevertheless, the future of such communities hangs by a thin thread. The arguments in favour of preventing depopulation (in order to preserve distinctive ways of life and to maintain attractive landscapes for the benefit of the urbanised population, according to McCleery (1979)) seem increasingly thin as economic recession prompts successive waves of fiscal retrenchment. There is no case on national economic grounds for giving priority to assistance in rural areas and small towns over urban areas, especially during periods of high unemployment. In the final analysis, therefore, the future of marginal communities is likely to depend on the extent of their electoral significance.

Migrant Labour in Host Communities. The residential segregation of foreign workers is an important dimension of urban structure in north-western Europe and is dealt with in Chapter 6. Here, attention is given to some of the broader social and political issues associated with the concentration of migrant labour in urban areas. The key to these issues lies in the type of work available to the immigrants. Since nearly all of the opportunities have involved low-wage jobs, immigrants have inevitably ended up in the worst housing and the most run-down neighbourhoods. Thus begins a spiral of social problems and conflicts (Leitner 1981).

The subordinate position of immigrants in the labour market is of course partly self-inflicted: for many the objective is to earn as much as possible as quickly as possible before returning home. The easiest strategy is to take on employment with an hourly wage where overtime and even a second job can be pursued. In any case, most immigrants do not have the skills or the qualifications to compete for better jobs. As Castles and Kosack (1973) point out, however, immigrants tend to be *kept* at the foot of the economic ladder by a combination of official restrictions and social discrimination. The net result is reflected by the statistics for West Germany in the mid-1970s: only one per cent of migrant workers held non-manual jobs (King 1976).

Similarly, immigrants' position in housing markets is partly self-inflicted: cheapness is of the essence. But the localisation of immigrants in camps, factory hostels, *hôtels meublés* (immigrant hotels), *bidonvilles* (suburban shanty towns) and inner-city tenements is also reinforced by restrictions and discrimination. Concentrated in such housing, immigrants find themselves in an environment which creates problems both for themselves and for the indigenous population. Trapped in limited niches of the housing stock, they are vulnerable to exploitation. A survey of housing conditions in the Ruhr, for example, found immigrant workers paying an average of 30 per cent more rent than German nationals, even though the latter had better accommodation (Böhning 1972). One response has been the notorious 'hotbed' arrangement, where two or even three workers on different shifts take turns at sleeping in the same bed. In France a more common response has been to retreat to the cardboard and corrugated iron bidonvilles, where although there may be no sanitary facilities immigrants can at least live cheaply amongst their fellow countrymen (Granotier 1970; Hervo and Charras 1971). In such circumstances, health and welfare-related problems proliferate, further polarising the immigrant community in terms of social well-being. Yet most foreign workers do not have the same civil rights as the indigenous population (Rose 1969); nor are local social services adequately geared to deal with their problems (De Haan 1976). Educational problems have become particularly acute as a second generation of immigrants makes its presence felt (Beyer 1980; Bilmen 1976). Already there are over 800,000 foreign children in the educational systems of both France and West Germany and more than 250,000 in Switzerland. In some cities, such children represent over 20 per cent of the total school population; in some school catchment areas they represent an overwhelming majority. Educational systems and educational resources have so far proved unable to deal with the special problems of language and culture faced by immigrant children. Moreover, educational issues have become a common focus for the mistrust and resentment felt by local residents towards immigrant communities (Kane 1978). Social conflict of this sort serves in turn to reinforce the separateness and isolation of immigrant communities, accentuating cultural and religious differences and preventing assimilation (Rist 1979; Trébous 1980; Wrochno-Stanke 1982).

Such segregation and subordination may also have a far-reaching influence on the overall social formation: indigenous workers, it is argued, finding themselves in a superior position to migrants, come to see some justification for hierarchical social organisation. The *status quo* is thereby legitimised (Carney 1976). At the same time, indigenous communities may feel their security threatened by the immigrants' 'invasion' of job markets, housing markets and welfare systems. This produces a general tension which has been transmitted into national political arenas by the likes of Enoch Powell (in the United Kingdom) and James Schwarzenbach (in Switzerland), and which occasionally transforms social conflict into violence at the local level (Herfurth and Hogweg-de-Haart 1982; Mayer 1975; Miller 1982). For the most part, however, the dominant migrant-host relationship is what Giner and Salcedo (1978) call 'passive subordination' — a lack of integration into the indigenous working classes combined with a general acceptance of the domination and authority of the social and institutional structures of the host community.

Source Areas of Migrant Labour. From 1945 until the late 1960s it was more or less taken for granted that the flows of migrant labour from Mediterranean countries to the industrial economies of northwestern Europe were beneficial to the areas of origin since they were all suffering acutely from unemployment and underemployment (see, for example, Lutz 1962). In addition, it was argued, the migrant-labour system brought benefits to source areas through remitted savings which provided valuable foreign exchange; and further benefits would accrue as returning migrants brought back modern work habits along with newly acquired industrial skills. Thus, the conventional view emerged that the recruitment of Mediterranean workers constituted a form of development aid (Beijer 1969; Hume 1973). During the 1970s, however, there emerged an alternative interpretation (Böhning 1975; Carney 1976; Nikolinakos 1973; Piore 1979; Schiller 1974). This perspective concedes that the migrant-labour system provides useful foreign exchange for Mediterranean countries over the short term, but challenges the view that it can enhance the prospects for development in source areas. In fact, some argue that the social and economic costs of labour emigration may outweigh the benefits, and that emigration constitutes 'a form of development aid given by poor countries to rich countries' (Castles and Kosack 1973: 8).

The weight of evidence from case studies of migrant labour certainly emphasises the failure of the system to promote development in the source areas. An extensive literature review by Swanson (1979) found several instances where productivity in the economic base of source areas was thought to have decreased as remittances from migrants had come to replace agriculture as the peasants' economic mainstay. For the most part, however, disappointment with the migrant-labour system stems from the failure of returning workers to exert any catalytic or beneficial effect on the economic health or social structure of their area of origin (King 1978). Several writers have suggested that, rather than constituting a flow of skilled, disciplined industrial workers with innovative ideas, liberal attitudes and sizeable savings to invest in new enterprise, returnees have predominantly been failures: people who were unable to adjust to industrial society (Böhning 1972; Kayser 1967; Trébous 1980). Failures or not, it is generally agreed that few returnees acquire skills or work experience that can be considered important to the home economy. Of the migrants who do obtain better jobs while living away, most are only semi-skilled, with little more than a brief training in how to carry out a specific operation in an assembly line (Lianos 1975; Paine 1974; Rhoades 1978, 1980). Moreover, returnees generally have little desire to continue in industrial employment upon return. The dream of most is to be independent and self-employed, which usually means setting up a small business such as a grocery shop or a taxi service.

The reality, however, is that relatively few returnees invest their savings – which are often quite considerable – in new enterprise. Nor do they invest their savings in agricultural land, having been 'spoiled' by urban lifestyles (Gmelch 1980). Rather, the most common form of investment is in housing. In southern Italy the explosion of migrant-financed building has been described as 'one of the most dramatic features of the changing rural landscape' (King 1977: 244). This certainly conveys an air of prosperity, but makes little contribution to the local economy apart from the temporary jobs generated in the building industry. Most of the savings which are not spent on housing appear to be splashed on highly depreciable consumer goods and luxury home furnishings. Rhoades (1978), describing the home of the typical returnee to Spain from West Germany, asserts that it is

> lavishly furnished and decorated with virtually everything the modern mass consumer markets offer. It is no exaggeration to define the situation as conspicuous consumption run amuck . . . Furnishings and interior decorating more often than not appear as Spanish replicas of German middle-class homes, an atmosphere dictated, for example, by *Quelle* catalogue, the German version of Sears or Spiegel. This fervour for imitation includes plush sofas, chandeliers, and wall-long buffets that serve as display cases for German china and glass articles. The built-in bar is always well-stocked with a multitude of liquors. Bathroom and kitchen walls are tiled and hallways are decorated with plaques from German cities, Black Forest cuckoo clocks, and other tourist paraphernalia. Electronic gadgets and appliances abound: televisions (some have two sets), stereos, radios, washing machines, cameras and watches. (p. 114)

Similarly, a survey of Greek returnees found that washing machines had been purchased by 89 per cent, televisions by 85 per cent, refrigerators by 77 per cent, electric kitchen ranges by 75 per cent, furniture by 71 per cent and stereo equipment by 59 per cent (Unger 1981). Indeed, the whole objective of temporary emigration seems to be strongly influenced by the idea of conspicuous consumption (Brettel 1979).

Nor are returnees very innovative beyond the level of superficial consumerism, Despite thinking of themselves as vehicles of social change and bearers of modern culture (Lopreato 1967), there is little evidence of change in the attitudes, values or social structure of source areas (Swanson 1979; Wiest 1979). Part of the reason for this, as Gilkey (1968) suggests, may be that returnees adopt a haughty demeanour which is resented by their countrymen. This, in turn, helps to explain some of the bitterness and frustration evident amongst returnees (Kenny 1972; Rhoades 1978). Such feelings are reinforced by the envy and suspicion of less prosperous neighbours; and they are often taken a stage further by a kind of 'reverse culture shock' (see, for example, Bernard and Ashton-Vouyoucalos 1976; Eikaas 1979; King 1977; Nicholson 1976). Thus, rather than acting as agents of innovation and catalysts for positive social change, returnees may have the effect of disrupting the social environment to the point where special regional policies aimed at re-integration are required (Signorelli 1980).

3 POLITICAL ECONOMY AND SPATIAL ORGANISATION

An understanding of Europe's territorial organisation, as Professor Mead emphasised in his presidential address to the Geographical Association (Mead 1982), is crucial to the 'discovery' of Europe and an appreciation of its geography. This chapter examines the European political quilt and deals with the changing interactions between territorial organisation, ethnography and modern political economy. The logical starting point for such a review is the system of nation states, which has been one of the fundamental bases for economic, political and social development within Western Europe. This, in turn, leads to a consideration of the implications of economic and political integration and of the significance of counter-trends of separatism and devolution. Finally, the issues raised by these trends are set in the broader context of European electoral systems, political parties, power structures and social movements.

The Rise of the Nation State

The nation state, according to Mead, 'is at once one of the most positive and most negative institutions that Europe has given to the world' (1982: 196). Within Western Europe the system of nation states, once established, fostered the economic, social and political organisation required by the industrial revolution. At the same time, strong competitiveness within the system provided an incentive to technological innovation (E.L. Jones 1981). On the other hand, nation states have fostered xenophobia and jingoism to the point of habit, encouraging political conflict and contributing to the subordination of minority cultures and ethnic groups.

It should be made clear that the nation state simply represents the conjunction of the state − 'a legal and political organisation' − and a nation − 'a community of people' (Seton-Watson 1977: 1). Put another way, the nation state represents an overlap between ethnicity and polity (Finer 1975; Grillo 1980). Such an overlap is nowadays taken for granted throughout much of Western Europe, but it was not until the late eighteenth century that the 'national identity' required for the creation of nation states began to emerge:

In the ferment of ideas first laid down in the eighteenth century, new concepts of community and identity arose. A vigorous interest in vernacular languages, in folk traditions, in the differentiation of types of man, strengthened by a widening horizon of literacy and the ability to disseminate knowledge through printing, all added to a new view of allegiance and community – nationalism. (Mellor and Smith 1979: 38)

Gradually, these feelings of common identity began to undermine the hegemony of the great European dynasties – the Bourbons, the Hapsburgs, the Hohenzollerns, the House of Savoy, and so on. After the French Revolution and the kaleidoscopic changes of the Napoleonic wars, Europe was reordered in 1815, to be set in a pattern that was to last for most of the next hundred years. Denmark, France, Portugal, Spain and the United Kingdom already existed as separate, independent nation states. The nineteenth century saw the unification of Italy and of Germany and the creation of Belgium, Greece, Luxembourg, the Netherlands and Switzerland as independent nation states. Early in the twentieth century they were joined by Norway, Sweden and then Finland. Austria was created in its present form in the aftermath of the First World War, as part of the carve-up of the German and Austro-Hungarian empires. At the same time, the victors took advantage of the opportunity to readjust their boundaries in the hope of tidying up the mosaic of ethnic sub-regions. In 1921 religious cleavages in Ireland resulted in the creation of the Roman Catholic Free State (later Republic), with the six Protestant counties of Ulster remaining in the United Kingdom. Finally, the rise and fall of the German Third Reich precipitated the eventual split of Germany into two states – East and West – and allowed further modifications to the boundaries of her neighbours as the Allied powers once again attempted to tidy up the map of Europe.

These developments reflect the long struggle to match polity to ethnicity, to make state boundaries fit populations with feelings of (or at least the potential for) common identity. As Fishman (1972), Knight (1982) and others have pointed out, this struggle involved two complementary processes: (i) states attempting to build 'nations' from a diversity of peoples, and (ii) peoples with a common identity ('nations') attempting to create an autonomous state. Grillo (1980) refers to these respectively as the 'ethnicisation of the polity' and the 'politicisation of ethnicity', and suggests that most European countries have experienced both processes in different ways and at different times. Most authorities would agree with Wallerstein, however, that 'the creation of strong states . . . was a historical prerequisite to the rise of nationalism' (1974: 145). That is, polity preceded ethnicity. It is also widely agreed that the fundamental logic for these trends towards national integration was provided by changes in economic and political organisation. Nation states, in other words, were constructed in order to clothe, and enclose, the developing political economy of industrial capitalism. They were not necessarily 'natural' entities which developed from distinctive cultural or philosophical bases (Miliband 1973; Poulantzas 1978; Tilly 1975; Tivey 1981). It follows that the process of building nation states involved the resolution of successive crises involving the interaction of territory, economy, culture and government (Dyson 1980; Knight 1982). The outcomes of these crises were fundamental not only to the political map of Europe but also to the whole character of its economic and social geography.

In order to establish the required feelings of common identity, states which sought to become nations — and nations which sought to mobilise their people in the demand for a state — had to engage in creating and diffusing a distinctive personality or character. Much of the ideology and symbolism of nation states has centred around the systematic mythologising of history, reinforced by the stereotyping of outsiders. An important outcome of this was the jingoism and xenophobia which set the context for the First World War, nurtured the ambitions of the Third Reich, and has hampered postwar attempts to integrate the states of Western Europe within common economic, cultural and political frameworks. Of more immediate importance to regional social geography have been the outcomes of processes whereby national identities were legitimised *within* states: the 'ethnicisation of the polity'. The development of 'national' identities and cultures has often meant the penetration of a territory and the absorption of its people by the culture and language of one powerful region or group. Economic differentials have been reinforced by educational systems, socialisation processes and legal codes, resulting in socio-cultural and political tensions between 'core regions' (or 'centres') and 'peripheries' (Giner and Archer 1978; Rokkan 1980; Rokkan and Urwin 1982; Shils 1981). Hechter (1975), likening the relationships between the dominant region and the dominated to those between the imperial state and its colonies, has argued that peripheral regions have been made dependent on core regions through what he calls a 'cultural division of labour'. Peripheral regions, in other words, are manipulated into specialising in the production of foodstuffs, raw materials and manufactures for which demand is relatively inelastic. The result, he argues, is a series of 'internal colonies' which are permanently disadvantaged by their subordinate economic relationship to the core region; they also suffer, like overseas colonies, from the cultural imperialism of the core region, sacrificing language and culture in the cause of efficient economic integration with the core. There are, however, few states in Western Europe where ethnicisation is complete, as Stephens (1976) has shown in his detailed account of linguistic minorities in Western Europe. Nation states generally crystallised outwards, by slow accumulation from central nuclei, so that surviving cultural minorities tend to be found in geographically peripheral locations or in extreme or inhospitable territory. 'Secondary nations' have therefore persisted as an important dimension of core-periphery relationships (Strayer 1963). Indeed, the nationalism of residual cultures has recently become a salient feature of the social geography of Western Europe (see pp. 66-72).

Social Well-being and the Changing Role of the State

'The proposition that industrialisation and urbanisation created the need for a rapid expansion of state functions', writes Elder (1979: 58), 'hardly requires elaboration'. The advent of the industrial revolution brought problems relating to public health, housing, civil disorder and social deprivation which could only be resolved by state intervention. In addition, the increasing personal incomes associated with industrialisation brought about a demand for public goods such as education. The central-place functions associated with urbanisation also impelled the growth of public expenditure (Tarschys 1975), while governments were forced to take on the role of managing the economy in order to provide a stable price system for the successful operation of private industry (Galbraith 1967). European history also shows that wars and the cyclical depressions of industrial capitalism drew state intervention further into welfare-related areas once the pattern had been established (Peacock and Wiseman 1967). Of course, the *capacity* of states to intervene in such matters was dependent on economic growth. It should be noted, however, that the changing role of the state did not follow directly in the wake of economic growth. The first major wave of industrialisation in France can be dated to the July Monarchy or the Second Empire (i.e. just before or just after the mid-nineteenth century); in Germany it was certainly no earlier. Yet it was imperial Germany that created the prototype of the welfare state, whereas the French had to wait until the Radicals' efforts between 1902 and 1914. Similarly, Britain was the first country to take off in terms of industrialisation and modern economic growth; yet by 1914 she had been overtaken in terms of welfare provision by Sweden, whose own industrial revolution did not begin until the last quarter of the nineteenth century. The reasons for this lack of synchronisation between economic growth and state activity lie in the complex arena of politics. The development of mass political movements was often a crucial trigger for increased state involvement and, as Elder (1979) points out, all of the modern political ideologies — conservatism, liberalism and socialism — have had a hand in the expansion of state activity in economic and social life. To some extent, the first two were impelled to this by the rising strength of the third. It is clear, however, that liberals and conservatives had their own reasons for indulging state intervention: the necessity for social control, for example, and the urge to tailor education to the needs of industry in the cause of 'efficiency'.

Trends in State Development. In detail, then, the development of state functions has been complex. It is possible, however, to identify some common trends in the nature of changes which have taken place. First among these, of course, has been the dramatic expansion of public expenditure as governments have become increasingly drawn into the creation of *welfare states* (Flora and Alber 1981). Italy, Sweden and the United Kingdom have already reached the stage where more than 50 per cent of their GNP is committed to public expenditure. Indeed, most Western democracies are approaching this figure, having almost doubled their share of the national economy since the 1950s (Bira 1971; Nutter 1978). It is important to remember, however, that there are significant differences in the *kind* of welfare state that has emerged in different countries: a critical point in terms of the eventual impact of state activity. Compare Italy and Sweden, for example. The roots of the welfare state in Italy can be traced back to the depression years of the 1930s when the Instituto per la Ricostruzione Industriale was set up. After the Second World War the expansion of the public sector took place for a variety of motives under conservative (Christian Democrat) governments. Major components of this expansion include the attempt to deal with underdevelopment in the south (through the Cassa per il Mezzogiorno), the provision of financial assistance to private firms in temporary difficulties (through agencies like the Gestioni e Partecipazione Industriale), the creation of public enterprises in order to break private monopolies (as in the cement industry), and the conservation of domestic jobs threatened by foreign competition. The 'profile' of the Italian welfare state is thus weighted somewhat towards regional economic development and employment protection, with health, education, pensions and social welfare assuming a secondary role. In Sweden the welfare state built up by successive socialist (Social Democrat) governments has involved little expansion of the boundaries of the public or nationalised sector of the economy. Instead, private industry has been encouraged to create economic growth, while the state has attempted to ensure an equitable distribution of the product through sophisticated welfare programmes. The Social Services Commission, for example, initiated radical health and unemployment schemes between 1938 and 1951, which were followed by the introduction of a pioneering pensions scheme which guarantees all retired workers an annual income roughly two-thirds of their earnings averaged over the best fifteen years of their working life. Both examples, however, point to an important corollary of the expansion of welfare states: the blurring of the boundary between the private and the public sectors. Government decision-making thus penetrates practically every branch of the economy, initiating a complex series of socio-geographical side effects (Dicken and Lloyd 1981; Maunder 1979; Sheahan 1976).

Another trend which is observable among Western European states is the expansion of their functions of *social control*. As Scase (1980) points out, this also stems ultimately from the processes of industrialisation and urbanisation: as traditional community and family structures are transformed in response to the needs of urban/industrial life, so traditional sources of control and support are attenuated. The state has stepped in — to an increasing degree — to replace them with formal agencies; not only the police but also social workers, for example, together with 'street-level bureaucrats' like housing officers and health service functionaries. Such developments have created some concern about the deterioration of civil rights and political freedoms, and have underlined the increasing contradiction in the role of the state as both 'protector' and 'despot'.

A third and very widespread trend among Western European countries has been the *centralisation* of the functions of the state. With the acquisition of larger and more complex roles in terms of welfare and social control, the overall tendency has been for local and regional activities to be 'rationalised' into centralised national bureaucracies. Centralisation can also be seen as a product of metropolitanisation, economic integration (at the national level) and socio-cultural homogenisation: government organisation has merely matched the changing scale of economic and social organisation. These changes, in turn, have been reinforced by technological changes in national communications and media systems (Sharpe 1979). Paradoxically, even regional economic and social policy-making has become centralised. France probably represents the extreme case of such centralisation (Birnbaum 1980; Gourevitch 1980; Graziano 1978); Italy, on the other hand, has been affected least of all by centralisation, the state being enmeshed in 'clientelism' at all levels: a pattern of relationships whereby local elites have come to act as patrons and mediators between government and society (Donolo 1980; Gourevitch 1980; Graziano 1978). Meanwhile, the power of politicians at both local and national levels has been constrained as power has been accumulated by the executive branches of bureaucracy. This, in turn, has led to crises in the legitimacy of political institutions, particularly in terms of traditional notions of representative democracy (Scase 1980). One response has been the intensification of demands for the devolution of power by the representatives of residual cultures in peripheral areas. Another has been the growth of forms of direct action, ranging from grass-roots pressure groups to urban terrorism.

Finally, the *autonomy* of even the strongest and most highly developed states has been *reduced* over the past fifteen to twenty years by the increasing internationalisation of capital (Blackborn 1982). In particular, the rise of the multinational enterprise has been seen as the source of a whole spectrum of problems:

> It fiddles its accounts. It avoids or evades its taxes. It rigs its intra-company transfer prices. It is run by foreigners, from decision centres thousands of miles away ... It overpays ... It competes unfairly with local firms ... It is an instrument of rich countries' imperialism ... It meddles. It bribes. It overturns economic policies. It plays off governments against each other to get the biggest investment incentives. (*The Economist* 24 January 1976)

As Williams (1979) points out, it is not only the host governments of multinational enterprises which are affected: the political balance and autonomy of 'base' governments has been gradually and indirectly affected by the existence of multinational enterprises. Of course, many European states – the Netherlands, the United Kingdom, West Germany and Switzerland, for example – are at once 'hosts' and 'bases'. Not surprisingly, it is difficult to pin down the influence of multinational enterprise on state activity. The whole relationship is further complicated by the fact that some multinationals are 'national multinationals' (the British-based Imperial Chemical Industries, for example) while others are 'international multinationals' (e.g. Unilever). As Schmitthoff (1973) observes, most European governments have fostered the development of the former, if only as a kind of 'national champion'; but it is not clear whether 'national multinationals' will in the long run pose less of a threat to the autonomy of the nation state.

What is clear is that states must increasingly compete with one another in order to attract capital. One important consequence of this has been the way in which states have amended the nature of regional policies in the hope of luring investment from the multinationals. Another has been the way in which states have approached social and economic reform: legislation designed to secure better working conditions, for example, may be abandoned or watered down for fear of 'frightening off' international capital. Most important of all, however, have been the attempts to regulate political and economic relationships with multinational enterprises through supra-national institutions such as the European Community.

The New Europe: Integration

After the Second World War a new dimension of territorial organisation began to emerge. A 'New Europe' developed around the framework of supranational organisations: a new political and economic order of 'immense' importance to the evolving human geography of the region (Blacksell 1981). The *political* roots of this shift towards integration are attributable to the outcome of the Second World War itself. The new global balance between the Soviet Union and the United States relegated Western Europe to a secondary place in world affairs. At the same time the divide which came to be known as the Iron Curtain quickly formalised the identity of Western Europe as a distinctive political arena. A further and related point is that within this arena there was a rough equality of power; there was no European 'victor' able to impose its hegemony or willing to give decisive leadership to other states. Western European leaders were quick to appreciate their subordinate position in world politics, and some saw that their main hope of self-assertion, even survival, lay in the integration of nation-states (Hoffmann 1982; Smith 1980). The *economic* roots of the new, integrated Europe, on the other hand, are generally attributed to economic imperatives stemming from the internationalisation of capital: the need for wider markets, the harnessing of international investment potential, the need for greater mobility of labour and capital, and so on. Having been created in response to the needs of industrial capitalism, the nation state found itself a 'contradiction' to the logic of post-industrial capitalism. In short, nation states required some form of integration in order to facilitate national economic development, while international business required nation states to integrate in order to increase the potential for the circulation and accumulation of capital (Kolinsky 1978; Miliband 1973).

Supranational Institutions

The oldest, most general and potentially most far-reaching component of supranational organisation in Western Europe is the Council of Europe. Set up in 1949, it now covers the whole of Western Europe. Its role is purely advisory, seeking to achieve 'a greater unity between its members for the purpose of safeguarding and realising the ideals which are their common heritage and facilitating their economic and social progress' (Council of Europe, quoted in Blacksell, 1981: 18). Nevertheless, it has been important in symbolising the idea that there is in Western Europe a common heritage, a common economic system and a distinctive social fabric. It has also played an important role in the harmonisation of standards of conduct, especially in the field of human rights. It has not, however, succeeded in taking the lead in establishing the institutions necessary for political and economic integration. Rather, these have emerged on a more pragmatic basis.

The Organisation for Economic Co-operation and Development (OECD), for instance, sprang from the insistence of the United States on guiding the postwar reconstruction of the West European economy. Beginning life as the Organisation for European Economic Co-operation (OEEC), the OECD eventually grew to embrace all of Western Europe together with Australia, Canada, Japan and the United States. The general objectives of the OECD have been to promote the expansion of world trade through the co-ordination of economic affairs. Its activity in the immediate postwar years in establishing a liberalised trading environment and free currency convertibility must be regarded as fundamental to the process of European integration (Blacksell 1981). Since the transition from the OEEC to the OECD in 1961, however, the focus of concern has been the co-ordination of economic relationships with the Third World. Scientific research, education and environmental protection have also been fostered within the membership, but not to the extent that it can be said to have influenced the nature or the pace of European integration.

The European Community also has its origins in a pragmatic response to the changed economic climate of postwar Europe. It was formed in 1967 by an amalgamation of three institutions which had been set up during the 1950s in order to promote progressive economic integration along particular lines: Euratom, the European Coal and Steel Community and the European Economic Community. The ultimate aim of the European Community is economic and political harmonisation within a single supranational government. Such objectives were perceived as an unjustifiable sacrifice of political independence (or, in some cases, neutrality) by Austria, Denmark, Iceland, Norway, Portugal, Sweden, Switzerland and the United Kingdom, who preferred to pursue the creation of a free-trade area without undermining the principle of national sovereignty. Hence the emergence of the European Free Trade Association (EFTA) in 1960. It has never developed a great deal of momentum, though it included over 100 million people in its heyday, and it undoubtedly helped lay the foundations of economic integration in Western Europe. Since the defection of Denmark and the United Kingdom to the Economic Community in 1972, however, it has been of minor importance.

The European Community, then, emerges as the major integrative force. Having expanded from its original six members – Belgium, France, Italy, Luxembourg, the Netherlands and West Germany – to include Denmark, the Republic of Ireland and the United Kingdom in 1972 and Greece in 1981, it now boasts a population of over 270 million, with a combined gross national product larger than that of the United States. It has developed into a sophisticated and powerful institution with a pervasive influence on patterns of economic and social well-being within its member states (Parker 1979; Vaughan 1979). The cornerstone of the Community is the deal worked out between the strongest two of the original six members. West Germany wanted a larger but protected market for its industrial goods; France wanted to continue to protect its large but highly inefficient agricultural sector from overseas competition. The result was the creation of a tariff-free market within the Community with a common external tariff for all imports from non-members, coupled with a Common Agricultural Policy (CAP) to protect the Community's agricultural sector. It is the CAP which dominates the Community budget, accounting for over 70 per cent of the total expenditure (Table 3.1). Its operation has made a significant impact on rural landscapes, the rural economy and rural levels of living. It has also influenced urban living through its effect on food prices, and in general terms it represents an impressive degree of integration of national policies. Nevertheless, it has been a consistent source of dissatisfaction, and since the accession of the United Kingdom to the Community it has become the source of serious disharmony. This largely follows from the United Kingdom's past policies on food, which were progressive, subsidising lower-income households. Accommodating the CAP meant a higher and regressive system of food prices without any compensatory benefits: peasant farming and inefficient agricultural systems had been purged from the economy long before, so there have been few beneficiaries of Community subsidies.

Table 3.1: European Community Budget Expenditures, 1980

	% of total
Agricultural guarantees	71.5
Agricultural guidance	2.3
Social	2.4
Regional	2.6
Research, energy, industry, transport	2.6
Co-operation	4.0
Staff administration, information, etc.	4.0
Repayments to members	5.0
Other	5.6

Source of data: European Parliament (1981).

The United Kingdom's position in the enlarged Community also highlighted the lopsidedness of Community policy in favour of rural areas compared with depressed industrial areas. As a result, the Community was persuaded to launch the Regional Development Fund from which to disburse investment grants to depressed areas of all kinds. The size of the fund remains limited, however, and the Community has so far been unsuccessful in developing a co-ordinated regional *policy* (see Chapter 5). Nevertheless, the Regional Development Fund does represent a further step towards the ultimate goal of complete economic and political integration. Other instruments which the Community has developed in pursuit of greater integration include the Social Fund (used mainly for retraining and redeploying redundant labour from the coal and steel industries), the Investment Bank, the Common Transport Policy and the European Monetary System (for details see Blacksell 1981; Collins 1983; Lee and Ogden 1976; Triffin 1980; Vaughan 1979). Recently, the most significant developments towards integration have occurred at the political level. The experience of a succession of European 'summit' meetings by heads of state has fostered a political maturity which has produced a genuine European dimension to global politics. The political base of the Community has also been strengthened substantially by the introduction of direct elections to the European Parliament in 1979. Although it remains relatively weak in terms of legislative powers, it has grown considerably in stature as an independent elected body with its own political identity and has helped to legitimise the Community by 'Europeanising' a whole spectrum of economic and social issues.

Problems of Integration. The road to integration is not a smooth one, however. Notable failures include the Community's Common Energy Policy and its attempts to hammer out the Common Fisheries Policy (Wise 1983). Economic integration has also generated some problems of its own, despite the rather rosy interpretations of Community analysts like Reichenbach (1980). Holland (1980), Lee (1976) and others have stressed that European integration has extended the processes of concentration and centralisation, creating both structural and spatial inequalities:

> the market of the Community is essentially a capitalist market, uncommon and unequal in the record of who gains what, where, why and when. Its mechanisms have already disintegrated major industries and regions in the Community and threaten to realise an inner and outer Europe of rich and poor countries. (Holland 1980: 8)

Slow-growing nation states such as the United Kingdom and Italy, in other words, are in danger of becoming backward problem regions within Community Europe (Kiljunen 1980). It has also been suggested that the probable enlargement of the Community to include Spain and Portugal along with the most recent member, Greece, will aggravate existing spatial disparities between nations and regions. The potentially massive costs of extending regional, social and agricultural policies would leave almost nothing for the poorer regions of the existing Community (Ball 1980) while their admission to the market area of the Community would seriously undercut the economic base of the Community's existing Mediterranean regions (Secchi 1982). Another negative aspect of integration and centralisation has been the inefficiency, inequity and alienation associated with the bureaucratisation of the Community. Fifty per cent of the funds earmarked for regional development, for example, are 'absorbed' by bureaucratic costs; technocratic decisions produce regressive subsidies and investment grants; and all the while 'harmonisation madness' diffuses the efforts of the

Community and makes for an increasingly bland and homogenised cultural environment (Dahrendorf 1972). These problems, together with renewed fears about national sovereignty, have prompted several member states to display an increasing resistance to the Community (Taylor 1982). In Greenland, which gained independence from Denmark in 1979, a referendum held in 1982 found 52 per cent of the voters (in a 75 per cent turnout) in favour of withdrawing from the Community. More serious, perhaps, was the decision of the British Labour Party in 1980 to make withdrawal from the Community part of its future electoral platform.

How 'European' are Europeans?

One interesting issue which follows from the economic and institutional integration of Western Europe concerns the perceptions held by Europeans of one another and of the idea of a mutual 'European' interest. There can be no doubt that the 'ethnicisation of the polity' has incubated a virulent collection of national prides and prejudices which are now immune to economic logic, political rhetoric and bureaucratic fiat. Germans will continue to be seen by most other Europeans as a little over-serious, preoccupied by work and inclined to arrogance; Scots will continue to labour under the popular image of a dour, unimaginative, ginger-haired people who love bagpipe and accordion music, dress up in kilts and sporrans, live on whisky and porridge, and generally spend as little as possible; Norwegians and Danes will continue to resent the Swedes' 'neutrality' during the Second World War; Italians will continue to think of themselves as the most gifted soccer players on earth; and so on. Often, of course, such caricatures have more than a little substance in the broad sweep of history. The two major facets of the stereotypical Frenchman — ardent lover and patriot/xenophobe — for example, can both be linked to the persistent national obsession with population growth discussed in Chapter 2. Similarly, the Germans are widely credited with having unswervingly worked their way to two world wars and two postwar economic miracles; the Italians have, after all, won the World Cup three times; and the Scots persist in propagating the image of eccentric Caledonianism under the guise of tourist information.

It is, of course, not quite as simple as these caricatures and generalisations imply. Nevertheless, it is true that national prejudices have persisted as a powerful and corrosive influence on attitudes towards integration. Kitzinger (1967) cites the public speech in 1962 of a former High Sheriff of Oxfordshire on the subject of the United Kingdom's potential partners in the European Community:

France: Our enemy since England became England ... A country which ... we have invariably defeated. A country which folded up and surrendered in 1940 without firing a shot or dropping a bomb ... the whorehouse of the world.

Germany: Only invented 100 years ago by Prince Bismarck. Has sought to devour the world ever since. Good soldiers but [they] follow bad leaders ...

Holland (*sic*): Only had a police force as an army ... Honest and slow, and eat far too much. Shatteringly divided by religion. Duller than ditchwater.

Italy: Clever artisans, clever artists, good waiters, pretty women, rather greasy little men terribly anxious not to be hit or kicked. In times of strife they betray their allies and run from their enemies. Scum of the earth.

Belgium: A little buffer state, divided into two parts who hate each other's guts. Apparently invented so that while Germany is invading it other countries can get ready for war ...

Luxembourg: ... about as big as Bedfordshire and not half as important ... (pp. 180-2)

Despite the prevalence of such insularity, many social scientists believed that elite opinion in general would favour European integration and that, furthermore, mass opinion would be 'pulled along' by elite attitudes. The first comprehensive evaluation of these ideas was carried out by Deutsch and his collaborators (Deutsch, Edinger, Macridis and Merritt 1967), who analysed commercial trade patterns, mail traffic, mass attitudes, elite attitudes and the content of the elite press. Their conclusion, that little or no progress had been made in the direction of Europeanisation, was substantiated by a series of parallel studies during the 1970s (Caporaso 1974; Feld and Wildgen 1976; Lindberg and Scheingold 1980); European institutions and leaders, it was reluctantly concluded, were unable to generate the 'charisma' necessary to overcome the inertia of past prejudice. A more optimistic evaluation was offered by Inglehart (1970a, 1970b, 1977), however, who suggested that a fundamental change towards a European orientation was taking place among the younger age groups of most nations. This revived the hypothesis that pro-Europeanism would increase, though now it was clear that mass opinion would have to be 'pushed along' by the attitudes of the young.

Attitudes Towards Integration: National and Regional Patterns. Since 1974 an extensive series of attitudinal data have been generated by the

Eurobarometer surveys commissioned by the European Community and made available for analysis by individual researchers. Seventeen large-scale surveys were undertaken in the eight years to 1982, and all of them included sections dealing with attitudes towards integration. Some questions have been included in almost all of the surveys. One of these was a simple question soliciting attitudes for or against the idea of European unification which had in fact been in widespread use in opinion polls since the early 1950s. It is therefore possible to piece together the changing pattern of support for European integration over a substantial period (Handley 1981). What emerges is that Belgians and West Germans have been the most staunchly and consistently 'European'. The French and Italians, meanwhile, started with lower levels of support, but have steadily increased them to match those of the Belgians and West Germans. The British and the Dutch, on the other hand, are alone in recording an overall decrease in their support for European integration; though the Danes, like the British (but not the Dutch) have occasionally had a majority of persons opposed to integration. It also emerges that the period 1970-3 was one of widespread disillusionment with the European ideal. After the expansion of the Community in 1972, however, there began an equally widespread upswing in support for integration which was significantly intensified by the direct elections to the European Parliament.

But does this mean that there has been a real increase in the mutuality of feeling between national groups? Further evidence from the Eurobarometer series suggests that there has been a general increase in the levels of trust accorded to one another by the inhabitants of different countries, though it is a relatively recent phenomenon and it is by no means universal. Drawing on data from surveys in 1970, 1976 and 1980, Handley (1981) shows that there was virtually no improvement in levels of mutual trust during the first half of the 1970s, apart from the improved status of Germans, (which continues a trend of rehabilitation into Europe which was clearly charted by Merritt (1968)). The Italians were least trusted by everyone; and they in turn recorded reciprocal feelings of distrust for the French, the Germans and the British. These feelings aside, however, most national groups accorded their European partners a positive, if modest, level of trust. The period between 1976 and 1980 saw a consolidation of these feelings, with the British, French and Italians benefiting most from the trend. Once again, however, the Italians' rating was consistently negative (even though they had evidently been converted to positive feelings towards everyone else). Another striking feature of Handley's results is the increasing

Figure 3.1: Attitudes to Integration in the European Community: Respondents who 'completely agreed' with the statement that 'In the European Community, a country like ours runs a risk of losing its own culture and individuality'

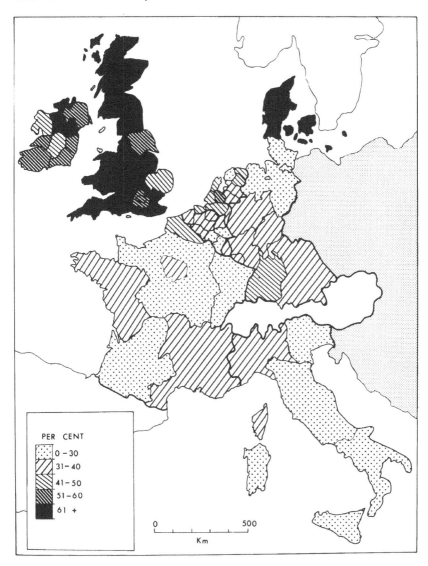

Source of data: Eurobarometer 10 (1979).

64   *Political Economy and Spatial Organisation*

Figure 3.2: Attitudes to Integration in the European Community: Respondents who indicated a willingness to make some personal sacrifice (e.g. paying more taxes) to help another Community country experiencing economic difficulties

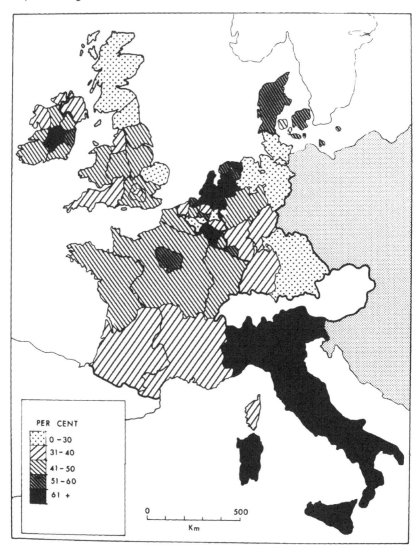

Source of data: Eurobarometer 10 (1979).

mutual trust between the original 'Big Two' of the Community: France and Germany. As Handley observes, one of the persistent themes running through the Eurobarometer data is the correlation between positive 'European' feelings and the length and intensity of a member state's involvement.

At the national level, then, the cumulative weight of evidence does tend to support the idea of an increasing European orientation among ordinary people. Integration, in other words, is not merely an economic and political phenomenon. The *regional* dimension of attitudes towards European integration is less well documented, though the Eurobarometer surveys can be disaggregated by region. Such an exercise produces some intriguing patterns. Figure 3.1 shows the 1979 response to the statement: 'In the European Community, a country like ours runs the risk of losing its own culture and individuality'. The first thing to note about this map is the generally low proportion of people agreeing that Community membership endangers national identity. The second is that the dominant patterns on the map represent national cleavages. Belgium, France, Italy, Luxembourg and West Germany are relatively homogeneous in their overwhelming approval of the Community, while the new members – Denmark, Ireland and the United Kingdom – are relatively homogeneous in containing a higher proportion of doubters. Both points lend further support to the idea of an increasingly Europeanised Community. On the other hand, Figure 3.2 suggests that most people are not yet willing to stretch their Europeanness to the point of personal sacrifice by paying more taxes to help solve the economic difficulties of others. This measure also produces a greater degree of regional variation within countries. The most striking features of Figure 3.2, though, are its lack of correlation with both Figure 3.1 and with the overall pattern of economic prosperity within the Community. Thus, we find a majority of Danes, for example, willing to make personal sacrifices even though between 48 and 52 per cent of all Danes perceive the system to be a threat to their national culture. On the other hand, the French, who do not see the Community as a particular threat, turn out to be cool or, at best, indifferent to the idea of personal sacrifice. These differences also cut across patterns of affluence: Germans are indisposed towards personal sacrifice whereas the equally prosperous Dutch are not. The only safe conclusion to draw is that attitudes towards European integration are still a long way from the kind of solidarity required for the 'ethnicisation of the polity' at the European level.

Decentralist Reaction: Separatism, Devolution and Reorganisation

While it is generally not questioned that integration and centralisation are the dominant trends in the economic, social and institutional organisation of Western Europe, it has been increasingly acknowledged that these trends have prompted a significant decentralist reaction in the *politics* of the region. As Tarrow (1978: 3) puts it, people are turning increasingly to local and regional institutions, 'reinforcing the territorial dimension in representation just as it is being displaced in policymaking and administration'. Sharpe (1979) suggests that this reaction has been expressed in three distinct movements: (i) regional ethnic separatism, (ii) local government reorganisation, and (iii) decentralisation of power to 'grass-roots' level through new institutions such as neighbourhood councils.

Regional Ethnic Separatism

Of all three decentralist trends, this was the least expected. Not only was the momentum of economic and political change aimed towards integration at both national and international levels, but technological changes were constantly expanding the potential for people to participate in ever wider communities of interest (Deutsch 1966). It therefore caused no little surprise and alarm when during the 1960s in prosperous Western Europe of all places – the cradle of the nation state – vigorous minorities presented themselves to astonished onlookers. Crystallising new political consciousness around issues of territory, language, history and 'way of life', separatist movements rapidly grew in size and influence. The phenomenon has been most pronounced in France, Spain and the United Kingdom, but it has affected almost the whole of Western Europe: only in West Germany has it been relatively unimportant.

A number of explanations have been advanced for this new – or at least revived – dimension of European social geography (Hechter and Levi 1979; Orridge and Williams 1982). One which seems at first to tie in well with the broad sweep of socio-geographic processes is the suggestion that regional and ethnic consciousness re-emerged as a result of the structural and spatial disparities which have been the inevitable by-product of modern capitalist development (Connor 1973; Mughan 1979; Nairn 1977). This explanation reaches its most sophisticated form in Hechter's (1975) model of internal colonialism in the United Kingdom, where capitalist development is seen as having exploited ethnic cleavages, thus fostering a sense of relative deprivation in the 'secondary nations' of the periphery. The sudden transition of Scottish and Welsh nationalism from tiny sects to more broadly based movements in the 1960s is explained by suggesting that this relative deprivation was bearable while the Scots and the Welsh could share in the domination of the British Commonwealth; but when this psychological compensation was removed by the postwar independence movement, regional ethnic consciousness became more acute. Yet although there is much that is attractive about the notion of uneven development precipitating ethnic consciousness, it is not adequate as an overall explanation for Western European ethnic separatism. It does not account for the separatist movements in relatively prosperous regions such as the Spanish Basque country, the French-speaking area of the Swiss Jura or the Flemish area of Belgium. Neither can it account for the timing of the resurgence of ethnic consciousness in regions which had never enjoyed the psychology of being part of a dominant imperial core (Birch 1978; Sharpe 1979; C. Williams 1980).

A more general and straightforward explanation of ethnic separatism is that it is simply a reaction to the increasing scale of human organisation. Thus, it is argued, people have become alienated from supranational, multinational and bureaucratised institutions, and have found in ethnicity and regional identity something to which it is more satisfying to relate and, more specifically, 'an all-purpose framework into which popular grievances can easily be fitted' (Philip 1980: 4). Moreover, in societies where consumption has become increasingly dominant, cultural difference becomes an item of value for its own sake. Similarly, separatist political rhetoric can be seen as a kind of self-indulgence, enjoyable by all, irrespective of status. Regional ethnic separatism thus has 'the potentiality of transforming failures into preferences, and inequality into pride' (Pizzorno 1963: 54). It has also been suggested that regional ethnic separatism springs from some kind of primordial sentiment which has finally been tapped by the advanced stages of democratisation achieved in Western Europe (Sharpe 1979). The timing of the resurgence of regional ethnic consciousness is thus seen as a product of the 'radical' socio-political climate of the 1960s, along with the resurgence of the women's movement, the emergence of an ecological movement, and so on. Again, however, the plausibility of the explanation must be qualified by the recognition of some awkward exceptions. Regional separatism in Spain, for example, can hardly be attributed to democratisation – a process which has only just begun after 40 years of harsh dictatorship.

Such exceptions to single-factor explanations tend to suggest that an understanding of regional ethnic separatism must be sought in the

combination and interaction of several contributory factors. This being
so, we should also consider several other influences which have
undoubtedly reinforced the process, but which are unlikely to have
been powerful enough to have rekindled regional ethnic consciousness
on their own. One is the role of historians and anthropologists (both
professional and amateur) in the discovery and reconstruction of
languages, cultures and 'ways of life'. Here is an important aspect of the
sociology of knowledge and its interaction with political economy
which has been almost entirely neglected. Nevertheless, it is clear that
academics and intellectuals have played an important part in the
elaboration and diffusion of what Grillo (1980) calls the 'myth history'
of ethnicity. Another influence concerns the effect of Third World
independence movements. As C. Williams (1980) points out, it seems
logical to expect that such movements were significant in stimulating
the political aspirations of ethnic minorities within the nation states of
the developed world. This is related to a further point: that support
organisations of the kind which have helped Third World countries to
establish themselves – the International Monetary Fund, the World
Bank, SEATO, etc. – also existed in Western Europe. Organisations
such as NATO, the European Community, the OECD, the Council of
Europe and the Nordic Council, it is argued, make separatism a more
secure and, therefore, more attractive proposition (Birch 1978). Finally,
it has also been suggested that the political leverage of separatist movements has been fostered by a variety of technological changes. These
include the impact of television news coverage and the vulnerability of
modern industrial technology and communications networks to
sabotage.

Whatever the relative importance of these various factors in shifting
the perceived balance of advantage from integration and assimilation
towards separatism and devolution, the outcome has been clear enough.
Regional and ethnic consciousness now represents a strong political
factor in many West European states. It has precipitated social disorder,
brought about changes in resource allocation and, in some cases, won a
considerable degree of autonomy. Figure 3.3 shows the major regional
ethnic groupings which have spawned separatist or devolutionary movements. The populations shown represent the total potential 'ethnic'
group. It should be noted that this is not necessarily an accurate reflection of the extent of ethnic consciousness. Most Scots and Welsh, for
example, share the same basic urban/industrial culture as the English,
and only a minority have ever voted for nationalist parties. Nevertheless, regional ethnic cleavages have often had an impact out of all

Figure 3.3: Minority Ethnic Sub-groups in Western Europe.

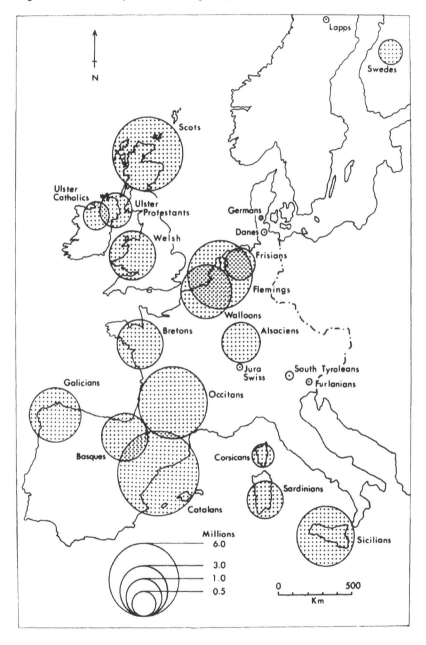

proportion to the numbers of people involved. Only four of the 37 West European ethnic groups identified by Krejci (1978), for instance, still have no kind of self-government: Alsaciens, Bretons, Corsicans and the Welsh. As Krejci discovered, however, it is impossible to generalise either about the nature of ethnic problems or about the degree of decentralisation achieved by separatist movements. It is appropriate, however, to outline the basic characteristics of each of the major regional ethnic groups.

Alsaciens. These are the approximately 1.3 million German-speaking inhabitants of the Alsace-Lorraine region in France. Several quasi-political associations have persisted with the objective of preserving the area's language and dialect. Minor outbreaks of violent protest have occurred since 1975, but no autonomy or institutional framework seems likely to be forthcoming. France is probably the least fertile environment as far as ethnic separatism is concerned, largely because of its particularly vigorous national ethnic consciousness. Moreover, the separatist movement in Alsace has suffered from resentment against the collaboration of leading autonomists during the Nazi regime. The bulk of Alsaciens have since preferred to run the risk of gallicisation rather than feel themselves completely isolated, an attitude which has been buttressed by the preservation of some freedom on local law and by the relative prosperity of the region (Gras 1982; Krejci 1978; Marcellesi 1979).

Basques. Of over 2 million inhabitants of northern Spain and south-western France, only 690,000 speak Euskera, the Basque language. While the French Basque region is an economically depressed area, the Spanish Basque provinces average 60 per cent above the national *per capita* income. The modern nationalist movement was launched in 1893, but it has been a new radical group, ETA (formed in 1959), which has done most to mobilise regional ethnic consciousness, often using terrorist tactics. After the demise of the Franco regime the Spanish Basque provinces were successful in achieving a considerable devolution of power from Madrid; French Basques, however, have no autonomy (Da Silva 1975; Heiberg 1982; Jacob 1975; Medhurst 1981).

Bretons. Of a population of nearly 2.4 million living in the Brittany region of France, about one million speak Breton, a distinctive dialect. Several clandestine paramilitary organisations exist, but they command a very limited following. No concessions have been made to decentralisation, though minor reforms have been undertaken in relation to the status of Breton as a school language (C. Williams 1980).

Catalans. There are between 6 million and 8 million inhabitants of Barcelona and its hinterland, the majority of whom speak Catalonian. The region enjoyed a brief period of home rule during the Spanish Republic in the late 1930s, before Franco came to power. Both leftist and rightist political parties later emerged with the principal aim of restoring regional political autonomy, which was eventually achieved in 1977 after the democratisation of the Spanish monarchy (Linz 1973; Payne 1971).

Corsicans. There are about 150,000 Corsican nationals in a total island population of 250,000. Ethnically more Italian than French, their consciousness was heightened by the immigration of French refugees from Algeria. Several small action groups have developed, and minor reforms in education have been achieved; but no significant degree of decentralisation.

Flemings and Walloons. These are the two major ethnic groups which formed the basis of the original bi-national state of Belgium in 1830. From the beginning, French (the language of the 4 million Walloons) has enjoyed higher prestige than the Dutch-based language of the 6 million Flemings, a cleavage which is reinforced by ideological, religious and economic differences. The failure of centralised institutions under a unitary government resulted in a revision of the Belgian constitution in 1970 and the subsequent introduction of a highly decentralised, federal system. Nevertheless, serious residual problems abound, including the position of the bi-lingual Brussels region, whose ethnic and economic interests are not compatible with those of either the Flemish or the Walloons (Heisler 1977; Stephenson 1972; Ter Hoeven 1978; Zolberg 1974).

Northern Irish. There are two basic groups, each of which is seen as a minority by the other. The numerical majority (1 million) within Northern Ireland are Protestants of English and Scottish descent. Their economic domination over, and discrimination against, the 540,000 Celtic-descended Catholics is seen by many as the root of the renewed support within the Catholic community for a united Ireland. All of the larger political parties are overtly sectarian, and both groups have spawned highly developed paramilitary movements and clandestine terrorist organisations. Since the intensification of the 'Troubles' in 1968 the spatial dimensions of the problem have become more pronounced, especially in Belfast, where residential segregation has produced clear, fixed boundaries between territorial heartlands. A series of attempts have been made to decentralise power to a Northern Irish parliament, but each has so far foundered on the unwillingness of the two groups to participate with one another (Boal and Douglas 1982; Fitzpatrick 1978; Hepburn 1980).

Scots. Of a total of more than 5 million, only 2 per cent speak Gaelic. The basis of ethnic consciousness, however, has always been politico-historical rather than language based. Scots have had a degree of autonomy ever since the Act of Union with England in 1707, and for a long time separatism was barely an issue of any significance. During the 1970s, however, separatism was the dominant theme in Scottish politics, with the Scottish National Party increasing its share of the vote from 11.5 per cent in 1970 to 30.4 per cent in 1974 under the slogan 'It's Scotland's Oil'. The movement has not sustained itself, however, perhaps because it soon became clear that the idea of 'Scottish oil' could never be much more than wishful thinking. In a referendum held at the end of the 1970s only a slim majority were in favour of an elected Scottish Assembly: not enough to justify decentralisation, according to the central government (Agnew 1978; Grasmuck 1980; Harvie 1977; Webb 1978).

Welsh. Of a total of 2.7 million, just under 20 per cent are Welsh speakers. Despite a more consistent history of separatist sentiment and a well-organised nationalist party (Plaid Cymru), the Welsh have never been able to mobilise ethnic consciousness to a very high level. At the peak in the mid-1970s, only three of the 37 Welsh parliamentary constituencies returned more than 30 per cent of the total votes cast to Plaid Cymru; and only 20 per cent voted for a Welsh Assembly in the 1979 referendum on devolution (Ragin 1979; G. Williams 1980).

Local Government Reorganisation

Local government reorganisation is not simply a reaction to the centralisation of power; it is also a reaction to changing patterns of settlement and the changing scale of organisation required for the efficient delivery of public services. It is, therefore, a highly complex subject. It must also be acknowledged that in some ways the reorganisation of local government is in fact a reflection of centralisation rather than a reaction to it. It is true, for example, that local government reorganisation has generally resulted in a transfer of some functions to central government, and has meant an increased dependence on central government for some aspects of finance (Bruun and Skovsgaard 1980; Kjellberg 1981; Martinotti 1981). On the other hand, it has also resulted in the widening of the scope of local functions, so that there is often no discernible change in the balance of centre-local power relationships (Dearlove 1973; Stanyer 1976). It is also relevant to note that, in terms of expenditure, local governments have significantly increased their relative importance over the last 20 years (Ashford 1978, 1979; Council of Europe 1975; Newton *et al.* 1980). Overall, then, there is a case for interpreting at least some aspects of local government reorganisation as an expression of decentralisation (Kalk 1971; Policy Studies Institute 1979). In this context it is ironic that it is the inertia of small-scale local government units – the *commune* or parish level – which is the target of much of the reorganisation in Western Europe. The traditions of the commune are almost everywhere very strong and as a consequence extremely resistant to change: in Belgium and Switzerland, for example, the rights of communes were constitutionally enshrined. The common reaction to nineteenth-century urbanisation was therefore to add a new tier of local government in an attempt to satisfy the needs of urban development while preserving the rural domain. Hence the county boroughs and counties of England and Wales and the *Kreise* and *Kreisfrei* of Germany. The twentieth century has brought new patterns of metropolitan development and new patterns of demand for public services, making both levels obsolete. More important, in the present context, it has also brought a new pattern of regional problems in the wake of the changing imperatives of economic development.

Reorganisation, then, has been motivated by several interlocking processes. It should not be surprising, therefore, that the *nature* of reorganisation is complex. It is, nevertheless, a widespread phenomenon, extending to most of Western Europe just as it has to most of the developed world. Within Western Europe the most extensive reorganisation has taken place in Denmark, Norway, Sweden and the United Kingdom. Root-and-branch reform is the exception, however, and in most countries reorganisation has been piecemeal and incremental: stop-gap devices have been used, *ad hoc* authorities have been created, inter-authority co-operation has been encouraged, and voluntary amalgamation has been made easier (see, for example, Gunlicks 1981; Muir 1983; Paddison 1983). If it is possible to generalise about the pattern of reorganisation, the first point to make is that the outcome is usually fewer, bigger units. Following the logic of metropolitan development and agglomeration economies, reorganisation has often been aimed at the territorial level of city regions. As the British experience illustrates, however, the vast political significance of territorial organisation means that both efficiency and equity are easily overtaken by partisan priorities. Nevertheless, the point to emphasise here is that most reorganisation exercises have provided an important counterpoint to the centripetal forces associated with central government. Moreover, there are some cases where reorganisation has in fact been an explicit reaction to centralisation. Here the underlying logic has stemmed from the unevenness of regional development and the consequent need for regional planning and regional administration.

There are two sides to this. One is the increased regional consciousness which derives from uneven development and results in the demand for some form of regional planning and/or administration. The other is the increased difficulty of managing national economic development without some form of regional framework with which to handle the problems of uneven development and 'unload' the regional co-ordination of public activity. Both have made for the introduction of a further tier of local government, though in some countries the solution has been cast in terms of non-elected bodies (such as the British and French economic-planning regions), resulting in the *deconcentration* of power rather than the decentralisation of power (Smith 1980). The Italian *Regioni* are generally regarded as the most comprehensive of the elected regional authorities within Western Europe. Postponed by successive Christian Democrat governments which were reluctant to have regional governments sniping at their authority, the 20 Regioni were eventually set up in 1970. They have quickly assumed a positive role in decentralising power, having acquired responsibility for implementing the laws and policies of the European Community and undertaken their own initiatives in international economic relationships in addition to their original constitutional powers to levy taxes and control primary industry, transport, tourism, police, housing and education (Condorelli 1979; Dente, Mayntz and Sharpe 1977; Evans 1981). This example, coupled with the devolutionary changes associated with regional ethnic separatism, has added to the momentum of the emerging regionalist movement in Europe (Urwin 1983). It must remain doubtful, however, whether other central governments will be willing to make such extensive concessions.

Decentralisation at 'Grass-roots' level

Reaction to the centralisation of society, the economy and the apparatus of state has been reflected at the 'grass-roots' level by the so-called neighbourhood-council movement. As Sharpe (1979) acknowledges, part of the movement's impetus undoubtedly stems from the popularly conceived inadequacies of local government, and the feeling that even local governments have become remote and insulated from community needs. He also points out that trends in European society towards higher levels of home ownership (over 35 per cent in the Netherlands, Sweden, Switzerland and West Germany, over 45 per cent in France and over 55 per cent in the United Kingdom) and extended education have made for higher levels of self-interest and a greater confidence and ability to participate in local affairs. Together, these forces have made the neighbourhood-council movement a very widespread phenomenon. The roots of the movement go back to the 1930s (Magnusson 1979), but it was not until the postwar period that it produced tangible results. One of the earliest examples is that of Rotterdam's 'urban district councils' set up in 1953 and composed of representatives elected by the city council with the objective of reflecting the social and political composition of each neighbourhood. By the 1960s such schemes were increasingly attractive as the problems of centralisation became more acute. A report by the Council of Europe (1966) on the development of local government observed that:

> The terms 'local democracy' and 'self government' can hardly be applied to municipal authorities where a hundred elected Councillors represent a million people ... The only remedy for this state of affairs is internal decentralisation within big towns and urban centres, the institution of an urban community pattern (p. 18)

By 1970 Rotterdam's scheme had been modified and extended to other large cities in the Netherlands. In Italy, Bologna had been a lone innovator in the 1950s and 1960s; but by the early 1970s there were 82 Italian cities with neighbourhood-council schemes (Dente and Regonini 1980). Norwegian cities also acquired neighbourhood councils during the 1970s (Kjellberg 1979), as did Swedish cities (SOU 1975). In England and Wales the neighbourhood-council movement got caught up in the debate over local government reform with the result that plans for a nationwide scheme were never implemented, leaving individual local authorities to develop schemes of their own (Hain 1976). In

Scotland, where local government reform was effected a little later, the neighbourhood-council idea was fully adopted, covering rural communities as well as city neighbourhoods (Martin 1981). The list is long, especially if it is extended to include the many spontaneous and *ad hoc* schemes involving local ombudsmen, neighbourhood troubleshooting centres, and so on. What remains to be seen, however, is whether the neighbourhood-council movement will in fact prove to be an effective centrifugal force. Some of the Swedish schemes have already collapsed, while the remainder have come to lead a quiet existence in the shadow of the parent local authority (SUO 1975). Moreover, it remains a matter of debate as to whether neighbourhood councils can really be made a focus for genuine public participation without degenerating into an elaborate exercise in public relations or lapsing into the hands of a self-selected knot of local *prominenti*. To put it another way, will the flow of information from communities to local authorities make government more efficient, sensitive and equitable, or will it merely serve as an early-warning intelligence system for more effective bureaucratic aggression and control (Knox 1978)?

Society and Politics in their Spatial Framework

Social Cleavages and Political Patterns

The political complexion of Western Europe is closely related to several dimensions of human geography in addition to the interactions of economic integration, ethnicity and government described above. The legacy of regional industrial development clearly relates to class-based political cleavages; settlement geography is reflected in rural-urban cleavages; and cultural geography often forms the basis of a sectionalism which extends well beyond regional ethnic consciousness (Lipset and Rokkan 1967). It is generally acknowledged that economic factors are very close to the centre of the web of influences on political loyalty, so that the class composition of particular areas is a prime determinant of political culture, party support and the structure of power. This relationship is well developed in Britain, where a high degree of industrialisation and urbanisation has resulted in pronounced class polarisation and the attenuation of other potential political cleavages. Typically, 60 to 70 per cent of the working-class vote goes to the Labour Party, compared with 10 to 15 per cent of the upper-middle-class vote. In West Germany the comparable figures are roughly 45 per cent and 10 per cent, and in France 60 per cent and 30 per cent (Rothman, Scarrow and Schain 1976). It is in Scandinavia, however, that class-based cleavages are most strikingly translated into voting patterns (Lijphart 1971). Nevertheless, nowhere in Europe can politics be read as a straightforward translation of class issues on to the political stage. Even in Britain, political parties cannot hope to achieve power on the basis of class loyalty alone; a certain amount of cross-voting has to be secured. The classic example, which has occupied a central position in the literature of political science, is the phenomenon of working-class conservatism (Mackenzie and Silver 1968; Nordlinger 1967; Parkin 1967).

While it should be acknowledged that such cross-voting is complex and only partly understood, it is clear that two of the dominant cleavages which sometimes cut across class divisions are those based on the rural-urban dimension and on religious affiliations (Rokkan 1970; Taylor and Johnston 1979). The conservative nature of most rural politics is well established. In France and West Germany, about half of the rural vote − peasants, farm workers and farmers − generally goes to right-of-centre parties, for example, while the rural dimension in British politics is overwhelmingly conservative. In Scandinavia there have emerged specifically 'agrarian' political parties, sharing neither in the radicalism of the left nor the conservatism of the 'bourgeois' city parties. Ties between rural and urban elites have remained few and tenuous, particularly in Norway, where the history of foreign (Danish) domination exacerbated the cultural bases of urban-rural differentiation (Rokkan 1967). Again, however, the patterns are by no means simple and straightforward. In Italy, for example, the rural-urban dimension is conditioned by regional variations in the system of landholding. The northeast is typified by small, independent farmers who traditionally lend strong support to the conservative Christian Democrats. In central Italy the injustices of the widespread *mezzadria* system of sharecropping results in support for the Communist Party and contributes to the long-standing communist control over local government. In the south high rates of emigration have hindered the emergence of a cohesive rural proletariat from the *latifundia* system of estate labouring, so that local notables continue to exercise a system of patronage, mainly through the Christian Democrat Party (Dogan 1967). Similarly, there is an important element of rural Communism in France, chiefly in the Massif Central and the areas of independent peasant farming in the south and southwest. Here, however, the communist vote is not so much related to landholding systems as to a tradition of general protest against the church, the state and government: any government (Frears 1978; Williams 1970).

Religious cleavages are more important than both class-based and rural/urban cleavages in much of Europe. According to Lijphart's (1971) analyses of voting patterns, religious cleavages are dominant in Austria, Belgium, France, Italy, the Netherlands, West Germany and Switzerland. This does not mean, except in Northern Ireland, that political issues are seen primarily in religious terms or that religious disputes are the main content of politics; rather it is that religious affiliation is an important variable in deciding the direction of a person's political outlook. It follows that religion may either reinforce or cut across class-based or other cleavages. In broad terms, the geography of religion results in three distinctive contexts: Catholic dominance, Catholic/Protestant parity, and Protestant dominance. It is in Catholic countries that religion has the most explicit political connection, for three reasons (Smith 1980): the wide claims and strong social doctrines of the church, its rigidity and flexibility, and the persistent priority of papal interests over those of the state. In Portugal and Spain the Catholic church has long been part of the ruling order and an important source of legitimacy for authoritarian rule. In Ireland, too, the Catholic church has encountered little by way of political opposition. In other Catholic-dominated countries (Austria, Belgium, France, Italy), however, a sharp polarisation has emerged between the church and secular society, with the priest and the schoolmaster often symbolising unrelenting combat. Anti-clericalism tends to reinforce anti-capitalism, and the two issues become interwoven. The church also becomes an object of attack because of its vast material wealth. More to the point, there is a clear geographical dimension to anti-clericalism: 'it is not just the disparities of wealth and their association with non-economic differences, but the occurrence of these in well-defined areas which make for the most implacable cleavages' (Smith 1980: 12).

In the Netherlands, Switzerland and West Germany there is parity between Catholics and Protestants. While religion no longer forms the basis for political or social discrimination, it does have a salience for politics, and the two churches remain strongly competitive. In the case of the Netherlands, sectionalism is generally agreed to have cut across horizontal divisions of class structure, thus taking the sting out of class conflict, at least until comparatively recently (Johnston, O'Neill and Taylor 1983). In all three countries, however, there is a clear trend towards the weakening of traditional correlations between religion and voting behaviour. In Britain and the Scandinavian countries, Protestantism dominates. Here, the 'national' character of Protestant churches has led to a ready identification with the state itself. They have therefore been a major pillar of establishment unity – to the extent that the

Church of England was once described as 'the Conservative Party at prayer'. Nevertheless, the more modest social role and more flexible social attitudes exhibited by Protestant churches have led to a weakening of the political significance of religion. As Butler and Stokes concluded in their detailed study of British voting: 'The ties of religion and party in the modern electorate are distinctly a thing of the past' (1969: 129).

The European Dictatorships. These socio-geographical bases of politics in Western Europe are for the most part articulated in party systems operating under a democratic framework. It should be noted, however, that three countries − Greece, Portugal and Spain − have only recently emerged from significant periods of authoritarian dictatorship. Greece is something of a special case, the 1967 military *coup* being a direct throwback to the right-wing dictatorships of the 1920s and 1930s. Under the pretext of forestalling a left-wing plot, the Greek military regime attempted to eradicate systematically the existing political framework under a repressive and arbitrary dictatorship. The sudden confrontation with Turkey over Cyprus in 1974, however, juxtaposed unpopularity with military weakness; the dictatorship caved in, and the 'despised politicians' were called in again. In both Portugal and Spain dictatorship was the product of what Trevor-Roper (1968) called 'clerical conservatism', the direct heir of the aristocratic conservatism which had been edged out by the liberal bourgeoisie in the late nineteenth century. Giner (1982) argues that the emergence of fascist dictatorships in all three Mediterranean peninsulas (Greece and Italy, as well as Portugal and Spain) in the 1920s and 1930s is attributable to the failure of the liberal bourgeoisie to cope with growing unrest from the peasantry/proletariat, to colonial failures overseas and to the sudden relative economic disadvantage created by industrial expansion in northwest Europe. The fascist dictatorships of Salazar (in Portugal) and Franco (in Spain) are thus to be seen as desperate episodes of modernisation rather than as symptoms of a recurring sickness: 'law-and-order militaristic or military governments consolidated iron-fisted reactionary coalitions and set about to complement with state intervention the endemic shortcomings of private capital formation' (Giner 1982: 175). By the 1970s the dictatorial formula had become exhausted through a combination of processes: secularisation, urbanisation, the expansion of the middle classes, and the international penetration of the economy. Democratic systems replaced dictatorship in Portugal and

Spain in 1974 and 1977 respectively, though both countries are still very much in transition.

Party Systems, Interest Groups and Social Movements. While the party systems and democratic frameworks of Greece, Portugal and Spain have yet to stabilise, those of the rest of Western Europe are well established. In most countries, present-day patterns can be traced back to the late nineteenth century when the extension of the franchise led to the development of working-class politics and the addition of social demo-cratic parties to those which had developed in response to the early cleavages of pre-industrial Europe: aristocratic/bourgeois, religious/ anti-clerical, rural/urban. This new class-based cleavage was further emphasised after the First World War by the emergence in many coun-tries of communist parties. Thereafter, and with the major exception of Fascism, the overall pattern of political parties has been remarkable for its constancy. The postwar period, in particular, has seen the line-up and relative strength of most parties change very little (Daalder and Mair 1983; Rose and Urwin 1970). The outcome in terms of the regional and national balance of power between parties is described below. It is important to bear in mind, however, that all of the liberal democracies in Western Europe have developed a series of institu-tionalised interest groups which are at once both complementary to and competitive with the party system in the way that they articulate social cleavages in the political arena. Typically, these include the trades union movement, the agricultural lobby, the churches, commercial and industrial confederations and professional organisations. There is con-siderable variation, however, in the role played by these interest groups in different countries and the degree to which they are 'house trained' by linking them to formal processes of government. It follows that their effectiveness in influencing regional and local issues will also vary a good deal (Philip 1982; Smith 1980).

It is also worth noting that during the late 1960s and early 1970s there emerged a variety of social movements which were characterised by their lack of association with the established political parties, the formal processes of government and the traditional, institutionalised interest groups. The most dramatic of these developments occurred as student/worker alliances during the spring of 1968 in Paris and West German cities and in the 'hot autumn' of 1969 in Italy. There followed a massive grass-roots political mobilisation across the whole of Western Europe, centring on issues relating to 'ecology', the rights of women, workers' democracy and, in particular, housing. Eventually spreading to

Portugal and Spain in the wake of the collapse of dictatorship, these social movements were first interpreted as symptomatic of an impending crisis of late capitalism in general and of democratic systems in particular (Castells, Cherie, Godard and Mehl 1974; Della Pergola 1975). Traditional political parties, it was argued, were proving unable to adjust to the new needs and attitudes of the working class, while centralist trends in economic and political organisation were creating increasing alienation (Smith 1976). The result was the proliferation of localised conflicts, squatters' movements and issue-oriented groups, whose energy seemed to offer a new way of mobilising working-class consciousness at national levels. The initial momentum of such social movements has not been sustained, however, and it has proved difficult to evaluate their real significance (Castells 1982; Ceccarelli 1982; Jensen and Simonsen 1981). Nevertheless, it is clear that social movements were not merely a by-product of the general social climate of protest during the late 1960s and early 1970s. They continue to exercise a political impact even though they are limited in size and strength. To a certain extent, their orientation has shifted to more general issues, as with the anti-nuclear movements in Britain, France, West Germany and the Netherlands, and the coalitions to support abortion in Italy. At the same time, though, new forms of local political mobilisation continue to emerge, such as the recent proliferation of housing struggles in the Netherlands and the spate of urban disorder in British cities in the summer of 1981 (Andereisen 1981; Draaisma and van Hoogstraten 1983; Rex 1982).

Electoral Systems, Electoral Bias and Electoral Representation. As Johnston (1979a) has shown, the nature of electoral systems has an important bearing on a broad spectrum of issues relating to territorial social justice. In Western Europe there has been a long history of concern about the sensitivity of electoral systems. The idea of equality of representation — that constituencies should be of equal size in terms of population — and the idea of proportional representation of political parties originated in the 1780s in Britain and France respectively. There was much debate, however, as to the way in which such democratic ideals could be translated into a workable electoral system. With the emergence of nation states and the extension of the franchise, the problem became pressing, and in 1864 and 1885 international conferences were held in Amsterdam and Antwerp in order to debate the relative merits of different systems (Carstairs 1980). Subsequently, most countries adopted one of the three main proportional systems:

the D'Hondt, the Hagenbach-Bischoff and the Sainte-Laguë methods. Only France and the United Kingdom have retained the less sensitive single-member system, whereby each constituency elects a single representative who is declared the winner by virtue of receiving the largest vote. West Germany falls half-way, with a mixed system; and it must be acknowledged that the proportional systems used in the rest of Western Europe vary a great deal in their detailed operation. Nevertheless, the generality of proportional systems is striking. Carstairs (1980) explains this as the result of pressure from religious, racial and linguistic minorities during the emergence of nation states. Electoral reform, in other words, was part of the price of the 'ethnicisation of the polity'. Proportional systems, however, do not necessarily eliminate electoral bias. O'Loughlin (1980a) has shown that certain social, political and regional groups are significantly disadvantaged in many Western European nations. His results confirm the insensitivity of single-member systems to voting patterns, but they also demonstrate that some proportional systems (particularly those of Switzerland, Norway, Finland, Italy, Ireland and Belgium) suffer from a significant degree of insensitivity. Moreover, all eleven of the European electoral systems in his survey were found to suffer from a considerable degree of malapportionment; that is, inequality in terms of the number of electors per representative. In general, rural constituencies are over-represented while large and growing metropolitan areas are under-represented. Most countries have attempted to correct such inequalities by redistricting, and O'Loughlin notes that Denmark, Sweden and Ireland have been able to effect substantial improvements. On the other hand, Norway and Switzerland are distinctive for their continued under-allocation of representatives to metropolitan areas and over-allocation of representatives to remote mountainous districts.

Because of the relative distribution of different population subgroups, variations in electoral representation also have implications for party support. In general, parties with strength in urban areas are disadvantaged while those with a base of rural support are at an advantage. Particular beneficiaries of this phenomenon have been the Gaullists in France, the Christian Democrats in Italy, the Fianna Fail in Ireland and the Christian Democrats and Christian Social Union in West Germany (O'Loughlin 1980a). In Britain it is inner-city areas which tend to be over-represented, so that it is the Labour Party which tends to gain most (Rowley 1975). As Taylor and Johnston (1979) point out, however, this particular bias is counterbalanced by the 'unintentional gerrymander' of Labour voters in the single-member system because of their concentration in 'safe' constituencies. This serves to emphasise the great complexity of electoral systems. Nevertheless, it is clear enough that bias and malapportionment are significant contributors to spatial variations in social well-being, even in proportional systems. As Knowles (1981) concludes from his analysis of malapportionment in Norway, the over-representation of rural areas has helped in the adoption of regional policies and transport strategies favourable to rural and coastal areas:

> These policies include bridge, road and ferry building, rural airport and airstrip construction, completion of the Nordland railway to Bodo ... retention of the express shipping route ... via many small ports, subsidisation of rural bus, rail, ferry and air services and special transport subsidies for North Norway ... Conversely the under-representation of Oslo and the Oslofjord area is likely to have been one reason for the delayed adoption of policies related to urban problems. Bus operating subsidies, for example, were denied to Oslo and Bergen until 1973. (p. 159)

Finally, it is worth noting the system of representation within the European Community, particularly since the overall trend towards political integration has been matched by an increasing redistribution of public funds by its central institutions. As with individual national electoral systems, the detailed operation of democratic representation within the Community is complex, and can only be touched on here. Johnston (1977) provides an analysis of representation within the Community's Council of Ministers, the European Assembly and the European Parliament before the advent of direct elections. He concluded that all three institutions were 'far from equitable'. In the Council of Ministers, for example, the distribution of votes is strongly biased in favour of the smaller countries. Germany, with twelve times the population of Denmark and twenty times the population of Iceland, is assigned only 3.3 times as many votes. As Johnston shows, this bias in favour of the smaller members is at least maintained and sometimes exacerbated when possible voting coalitions are taken into account. The European Assembly, which was the forerunner of the Parliament, also exhibited a clear bias in favour of the smaller members while France, Italy, West Germany and the United Kingdom were all greatly under-represented. When the relative power of each country's votes was taken into consideration (in terms of their potential contribution to different combinations of coalitions), the most favoured countries were Belgium,

Luxembourg and the Netherlands and the weakest were Denmark and Ireland. Although the system was revised for the directly elected Parliament, the new arrangements also suffer from malapportionment and bias: the relative strength of different countries has simply been shuffled a little. Luxembourg, with six members of parliament, has a ratio of one representative per 60,000 residents; West Germany has one per 750,000 residents. Once again, the relative power of each country's representatives depends on the possibility of coalitions, and this time it emerges that the Benelux countries remain the most clearly favoured while it is the 'Big Four' which are at the greatest disadvantage. It must also be recognised that the electoral system used for the direct elections to the Parliament varies from one member state to another. Each is allowed to select its own system and, not surprisingly, the choices tend to follow the systems used for national parliamentary elections. Britain, for example, has persisted with the single-member system, with the result that none of the minority parties, including the Liberals, won a seat in the 1979 elections (Lodge and Herman 1982).

Voting Patterns and the Balance of Power

The formal political complexion of Western Europe is constantly changing and so difficult to generalise. International comparisons of voting behaviour are in any case hampered by problems of comparability between parties: it is often a matter of debate as to which parties represent the 'left', the 'centre' or the 'right' (Stammen 1980). Indeed, the individuality of political parties has been highlighted within the European Parliament, where the socialist parties are the only ones with any degree of international coherence; other alliances have emerged as marriages of convenience rather than on the basis of strong ideological commitments. The Socialists also happened to be the largest single group to emerge from the first direct elections in 1979, although the Christian Democratic group is almost as strong and the Conservative, Communist and Liberal groups are sufficiently large to be able to tip the balance of power. At present, however, the European Parliament remains in its infancy, and no clear pattern of cross-national voting patterns has emerged.

At the *national* level, most political systems have matured to the stage where it is at least possible to identify the balance of power which arises from (i) the number of competing parties sustained by different political cleavages, (ii) their relative size, and (iii) the feasibility of coalitions between them. Following Smith (1980) it is possible to recognise three kinds of situation: imbalanced, diffused and balanced outcomes. It is unusual for national politics in Western Europe to be imbalanced to the extent that a single party has been wholly dominant for more than ten or fifteen years (Maguire 1983; Pedersen 1983). There are several instances, however, where a particular party has been indispensable for the formation of government even if it has not always held an absolute majority. Such a position has been enjoyed by the Italian Christian Democrats, the Irish Fianna Fail and the Social-Christians of Luxembourg for most of the postwar period. The French Gaullists, the West German Christian Democrats, the Austrian Socialists and the Norwegian Labour Party have also enjoyed a modified form of dominance for extended periods during the past forty years. Diffused systems are characteristic of Belgium, Denmark, Finland, the Netherlands and Switzerland. They each involve a relatively large number of parties, none of which has a decisive superiority, but several of which have a comparable share of the vote. Inevitably, these systems are prone to governmental instability. In the Netherlands, for example, the average life of postwar governments has been no more than two years. By and large, however, the parties involved in diffused systems have learned to co-operate with one another if only to avoid getting a bad name with the electorate.

The major requirement for a balanced outcome is that there should be a dominant cleavage within the electorate which is reflected by two major parties or by two clusters of parties. In fully balanced systems the two competing camps should each command about the same number of seats and should between them account for about 90 per cent of the total number of available seats. This situation is most nearly approached in Austria, West Germany and the United Kingdom, and the typical outcome over the long term has been for an alternation of government. Smith (1980) argues that within Western Europe as a whole there has been something of a trend away from diffused and imbalanced systems towards balanced systems, mainly as a result of the 'repair' of fragmented parties representing the broad left or the broad right. France illustrates the trend well. Diffusion and instability preceded a long period of Gaullist dominance. The emergence of a moderate bourgeois party was a novel departure in France, and for a while the spectacular rise of the Gaullist party suppressed the development of others. Only after a 'long march' (Johnson 1981) did the Socialists establish a rough parity with the Gaullists, eventually winning the national election of 1981. Not many electoral histories are so straightforward, though there has certainly been a general reduction in the prevalence of coalition administrations since the 1960s. It is also interesting to note that swings in voting patterns appear to be roughly synchronised across Western Europe, perhaps reflecting common responses to the broad ups and downs of the international economy. The international recession of the late 1970s, for example, was accompanied by a widespread swing in favour of right-of-centre parties; by 1979 only Austria, Norway and West Germany had left-wing governments (Sallnow and John 1982).

Whereas national governments come and go, the *regional* pattern of voting gives some idea of the underlying pattern of political allegiances. Accepting that left-right cleavages represent the broadest common denominator and that left-of-centre parties represent the most coherent international grouping, it follows that the most meaningful single picture of the political landscape is one which is cast in terms of left-of-centre voting. Figure 3.4 provides a general indication of the geographical bases of left-of-centre support at a time when the overall trend was to the right. Given the complexity and variability of party systems, it would clearly be unwise to lean too heavily on such a map. Nevertheless, it does illustrate several of the key dimensions of European electoral geography: the classic bases of left-wing support in industrial Britain, northern Germany, central Italy and southern Spain, for example, together with the broad geographical basis of support for socialism in Denmark, Norway and Sweden. Conversely, the likes of Switzerland and Ireland are notable for the general absence of support for left-wing parties beyond the immediate region of their metropolitan cores. It must be acknowledged, however, that regional gradients in party allegiance are relatively low. Rose and Urwin (1975), who analysed the degree of regionally based party support in 13 European nations, found that voting patterns had become increasingly 'nationalised' between 1945 and 1970 in every country except Finland, where regional voting had become more pronounced. This conforms with Rokkan's (1970, 1980) model of political development wherein territorial conflicts are succeeded by conflicts emanating from religion, economic sectors (the rural/urban cleavage) and class structure. But while regional gradients are diminishing it is true that some parties depend more on regional bases than others. Rose and Urwin found that, in general, higher levels of regional voting are associated with communist and conservative parties; socialists, liberals and centre parties rely more on national cleavages and national swings. Not surprisingly, the agrarian parties of Scandinavia stand out with some of the highest levels of regional voting, while the Dutch Catholic party and the Swiss conservative party are also associated with distinctive regional votes.

Figure 3.4: The European Left: Generalised Map of Constituencies Held by Left-of-centre Parties in National Elections, *c*.1980

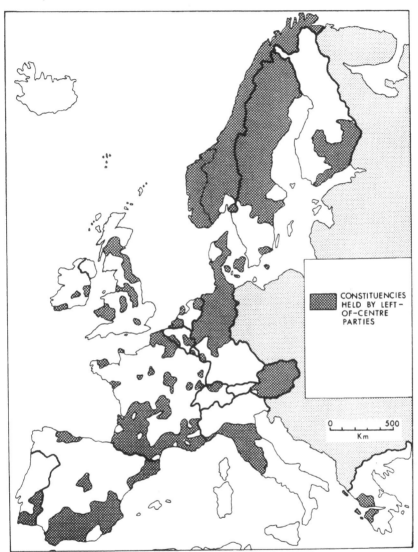

CONSTITUENCIES HELD BY LEFT-OF-CENTRE PARTIES

0   500
Km

Source: Compiled from data in Sallnow and John (1982).

4 RURAL EUROPE

Despite its contracting importance in terms of population and employment, the rural dimension of Western Europe remains essential to the character of regional landscapes, cultures, politics and settlement patterns. Agriculture, in turn, remains central to patterns of rural development. In the first instance, of course, the geography of agriculture has been conditioned by the physical environment and by the different systems of farming and landholding which have evolved in different parts of Europe. It has also been conditioned by the long-term consequences of the Industrial Revolution: the loss of rural handicrafts, an increasing specialisation in commercial agriculture and an increasing dominance of the rural economy by farmers, their labourers and their families (Mauret 1974). The result is a series of crop-regions, field systems and settlement patterns which are the classical discriminants of rural regions within Western Europe. The greatest influences on the geography of agriculture in the postwar period, however, have been (a) the structural reorganisation of agriculture, (b) the increasing intervention of the state in agriculture, and (c) the general 'revolution' in transport and mobility. Together, these changes have brought several new dimensions to the contemporary geography of rural Europe.

Aspects of Change in Rural Europe

In general, structural change in agriculture has involved five major trends:

(1) an increase in large-scale, specialised farm units;
(2) increased mechanisation (and therefore decreased levels of demand for labour);
(3) increased use of biochemical inputs;
(4) increased regional specialisation;
(5) an increased level of food processing and inter-regional marketing.
(Buttel 1980; Tracy 1982)

These trends have steadily increased the 'viability threshold' of farms, resulting in a widespread movement towards farm amalgamation.

Nevertheless, there remain a considerable number of economically marginal small farm units (Bowler 1983; Bryant 1974). Even in the highly rationalised and economically efficient agricultural sector of England and Wales more than 60 per cent of the farm units are so small as to be commercially marginal (Hirsch and Maunder 1978). Moreover, it has been the larger farm units which have been most successful in raising the capital necessary to take advantage of increased specialisation, mechanisation and biochemical inputs. As a result there has been a polarisation of affluence within the agricultural sector, reflecting what Commins (1980) describes as the 'dualism' of contemporary agriculture: big, prosperous units on the one hand; small, residual and increasingly marginal units on the other. Buttel (1980) suggests that the state has been drawn more and more into the agricultural sector as a direct result of this imbalance. Yet, he argues, although state intervention has generally been presented as an attempt to bolster the 'family farm', the net outcome has often been the subsidisation of larger producers, thus reinforcing the dualistic structure of the agricultural economy.

Meanwhile, the dominance of agricultural employment in rural communities has been dramatically eroded. As mechanisation and rationalisation have reduced employment opportunities in agriculture, the mobility revolution associated with increasing levels of car ownership has allowed an increasingly large 'adventitious' population to penetrate rural areas, generating a certain amount of employment, especially in service industries. Improved transport and communications have also helped to foster the re-emergence of rural manufacturing, while employment in tourism similarly owes much to postwar changes in personal mobility. Although these new sources of employment have rarely been sufficient to counter the loss of jobs in agriculture, the economic diversification of rural communities represents a major feature of postwar Europe. The social structure of rural communities almost everywhere has also been affected as the internal combustion engine has ended their isolation (Veldman 1981), and as structural change in agriculture has ended the traditional relationships between the different components of the farming community (Goldschmidt 1978; Münzer 1983). Within particular communities, however, the mobility revolution has by no means been universal in its impact. Thus, along with the overall attenuation of tight-knit, face-to-face, *gemeinschaft*-type community organisation there has appeared an important new dimension of rural society: the contrast between the 'transport poor' and the 'transport rich'.

Jobs in the Countryside

Perhaps the most striking outcome of these trends has been the changing pattern of employment opportunities in rural areas. Figure 4.1 illustrates the overall relative change in agricultural employment in Western Europe since 1950. Everywhere there have been significant reductions. Whereas a majority of Western European countries had more than a quarter of their labour force in agriculture in 1950, only Greece and Portugal had more than 20 per cent of their labour force in agriculture in 1980. In Denmark, France, Sweden, Switzerland and West Germany agriculture now accounts for less than 10 per cent of the labour force; and in Belgium, the Netherlands and the United Kingdom the figure is well under 5 per cent. The consequences of such trends for rural labour markets have been profound (Hodge and Whitby 1981; Marquand 1980). The point which should be emphasised here is that the contraction of employment opportunities in agriculture is directly related to patterns of rural depopulation which, in turn, have been central to the downward spiral of rural deprivation experienced by many rural areas (see below, pp. 110-13). Moreover, the loss of jobs in agriculture has been compounded, in many regions, by proportional losses from forestry and other primary activities.

Part of the resulting vacuum in job opportunities has been filled in *some* areas by an expansion of employment in other sectors. Rural manufacturing industry has steadily expanded its labour force during the postwar period, partly because of the effects of transport technology in 'shrinking' distance in terms of costs (Cawley 1979). Recently, however, some rural areas have experienced a very striking expansion of manufacturing jobs, prompting considerable debate as to whether the underlying causes are related to regional policies, the locational preferences of entrepreneurs, the agglomeration diseconomies of large cities or a more general restructuring of industrial capitalism (Urry 1984). Jobs in service industries have also expanded in many rural areas since 1945, mainly in response to the demands of the new adventitious population of commuters and retirees, but also as part of the expansion of public expenditure on administration and social services and, indeed, of the overall expansion in services such as banking and transport. Such jobs, however, are particularly susceptible to the displacement of labour by modern technology and by cut-backs in public expenditure (Dower 1980).

For some rural areas then (especially in the remoter, peripheral regions), tourism represents the only significant source of new employment. Indeed, there are some areas where the tourist industry has become the mainstay of the economy: Cornwall and the Lake District in England; large sections of the French, Italian and Swiss Alps; the Greek Islands; the southern coasts of Spain; and so on. Elsewhere, as in Jutland, the Highlands and Islands of Scotland and southern Italy, tourism represents a welcome addition to otherwise diminishing sources of employment (Gonen 1981; White 1976). It has long been recognised, however, that tourism is a double-edged sword as far as rural development is concerned (Bonneau 1979; Young 1973). Not only does it involve the socially and culturally disruptive presence of 'outsiders' (Greenwood 1972; Kariel and Kariel 1982), it also tends to dislocate 'traditional' industries, to inflate land and property values (thus edging the indigenous population from the housing market), and to 'deskill' the labour force. Seasonal unemployment and vulnerability to economic recession must also be set against the benefits of the employment opportunities (and spending power) created by tourism. As a result, there is growing support for the view that the net long-term effects of tourism may in fact tend to weaken the position of rural communities, increasing their dependency on metropolitan core areas and, ultimately, leading to a shrinkage in their employment base (see, for example, Vincent 1978).

The Rise of Agribusiness

Implicit in the structural changes affecting agriculture has been a general shift from 'farming as a way of life' to 'farming as a business'. Rather surprisingly, very little of the literature on rural Europe is devoted to capitalist farming. It is clear that, especially since about 1970, there has been a large increase in corporate involvement in agriculture. One aspect of this is simply corporate investment in agricultural land as a hedge against inflation or as a 'tax break'. Little is yet known, however, about the implications of such activity for the structure of rural society in general or the local rural community in particular (Massey and Catalano 1978; Newby 1980). Direct corporate involvement in agriculture itself – agribusiness – has been an inevitable outcome of the logic of specialisation and economies of scale (Mottura and Pugliese 1980). With greater specialisation, farms become less autonomous and self-contained as productive units, making for the penetration of an integrated, corporate system of food processing and distribution:

Agriculture has become increasingly drawn into a food-producing complex whose limits lie well beyond farming itself, a complex of agro-chemical, engineering, processing, marketing and distribution industries which are involved both in the supply of farming inputs and in the forward marketing of farm produce. (Newby 1980: 61)

Figure 4.1: Percentage of the Total Labour Force Employed in Agriculture, 1950-80

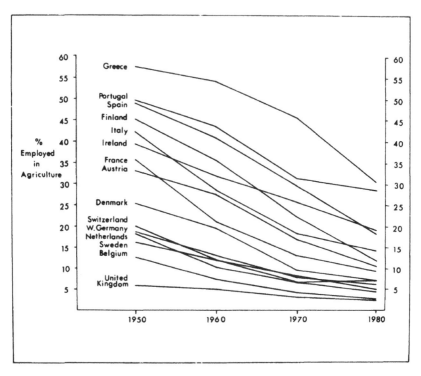

Source of data: National statistical abstracts.

It is in the actions of food-processing conglomerates like Associated British Foods, Nestlé and Rank-Hovis-Macdougall, Newby suggests, 'that the shape of agriculture and ultimately of rural society in virtually all advanced industrial societies is decided' (p. 62). The most common form of corporate involvement in agriculture centres on the forward contracting of produce at a fixed price. This not only weakens the independence of farmers, but also tends to transfer income from farmers and rural communities to the processing industry. Forward-contracting arrangements also reinforce the overall structural changes affecting European agriculture:

They encourage both fewer, larger holdings and increased specialisation so that the size of individual enterprises can be enlarged to fully achieve the prevailing scale economies. This trend ... is likely to lead to both a reduction in the numbers employed in agriculture, and a decline in the managerial role of those farmers remaining ... leaving them caretaker functions. (Metcalf 1969: 104)

Rural landscapes have also been affected as the logic of industrial production has been applied to agriculture (Thieme 1983; van Welsenes 1979). In northwestern Europe field systems have been rationalised, hedgerows and dykes removed, and mechanisation has virtually eliminated the need for gang labour, leaving the fields of most farms devoid of human life for most of the year. Factory farming has brought poultry and pigs indoors permanently, while many cattle spend their winter months indoors, and there are now 'zero grazing' techniques which may see them inside the year round. Only sheep steadfastly refuse to acknowledge the laws of industrial production, stubbornly refusing to prosper in regimented and sanitized conditions.

The 'Theft' of the Countryside and the Politics of the Environment

Such landscape changes are part of a broader assault on rural ecology associated with structural change in agriculture and, in particular, the side effects of agribusiness. Increased specialisation has reduced the practice of crop rotation and resulted in increased vulnerability to pests and disease, leading in turn to a heavy reliance on pesticides and herbicides. At the same time, specialised crop farms have come to rely on chemical fertilisers to replenish soil fertility. This increase in biochemical inputs has turned agriculture into a major consumer of energy – to the extent that European food production now requires as much energy per worker as manufacturing industry (Leach 1975) – and has intensified problems of soil and water contamination. Meanwhile, state intervention in agriculture has also resulted, indirectly, in adverse changes in rural environments. National and European Community price-support schemes have been pegged so high that it has been profitable for farmers to plough up almost any kind of uncultivated land in order to increase their output. Capital grants aimed at increased output have also resulted in the removal of hedges, woods, rough grasslands, streams and ponds, thus helping to create 'prairie' landscapes and destroy the habitat of wildlife. An increasing awareness of these issues, heightened by books such as *Silent Spring* (Carson 1969) and *The Theft of the Countryside* (Shoard 1980), has turned the rural environment into a serious political issue in most of northwestern Europe and parts of the Mediterranean (Ganser 1982; Lowe and Goyder 1983; Trilling 1981).

Moreover, conflict over ecological issues has been compounded by the simultaneous development of serious conflicts over access to the countryside. Under modern farming practices, livestock has become more vulnerable to diseases which are easily spread by humans; and farmers who use large quantities of toxic chemicals are naturally nervous about the presence of visitors to the countryside. As Newby (1979) points out, the simplest solution is to try and exclude outsiders altogether:

> Up go the 'Trespassers Will Be Prosecuted' and 'Keep Out' signs, the extra barbed wire is ordered and the shotgun is cleaned and prepared. Those who dare venture off public rights of way find themselves harassed and evicted. As a further disincentive footpaths may suddenly disappear under the plough or new fences are erected across them, signposts mysteriously vanish and perhaps an unsociable bull is (illegally) posted in a field traversed by a path. Once again the battle lines are drawn. (p. 221)

The ensuing conflict has been particularly sharp around the more heavily urbanised parts of Europe where the pressure of recreational visits to the countryside is greatest (see, for example, Tschudi 1979). Here, environmentalists and ecological pressure groups have joined in uneasy alliances with agricultural interests in order to designate areas of restricted access. This demonstrates that alliances formed in the conflict over access to the countryside are not necessarily the same as those which occur over issues of wildlife preservation and landscape change. Indeed, the line-up of protagonists in access conflicts will depend very much on the local context. In the marginal farming areas of upland Europe, where revenues from tourism are an important part of the rural economy, farmers may welcome recreational development while environmentalists resolutely oppose it. In general, however, it is the middle classes of urban origin who are most conspicuously involved in all these conflicts (Mormont 1983). Having 'claimed' the countryside — either as adventitious colonists or as intellectual guardians *in absentia* of arcadian virtues — the middle classes tend to be reluctant to share it with others: either capitalist farmers or the increasingly mobile but 'philistine' working classes.

Settlement and Society in Contemporary Rural Europe

What does all this mean in terms of patterns of rural development, settlement and social well-being? It must be acknowledged at the out-set that the constraints of climate, soils, altitude and aspect make for a certain degree of continuity in the rural geography of Western Europe. Although it should be stressed that marked variations take place over quite short distances, it is useful to think in terms of five main agricultural environments. In Scandinavia, with its short growing season and cold, acid soils, oats, rye, potatoes and flax are the major crops, with hay for cattle. The more humid and temperate regions of 'Atlantic' Europe (Belgium, France, Ireland, the Netherlands and the United Kingdom) are dominated by dairy farming on meadowland, sheep farming on exposed uplands and arable farming (mainly wheat, oats, potatoes and barley) on drier lowland areas. Central Europe is characterised by grains and root crops; Mediterranean Europe is characterised by tree crops, hard wheat and pastoralism; while mountain regions throughout Europe shares a dependence on pastoralism eked out with arable produce from valley floors. Within this broad framework there are numerous specialist farming regions where agricultural traditions have influenced local ways of life to produce distinctive socio-cultural environments. Examples include classic viticultural regions such as the middle Mosel, Burgundy and Bordeaux, and the fruit and vegetable *huertas* of eastern Spain – all areas where the structure and rhythm of rural society remains intimately meshed with the local economic order.

It must also be acknowledged that there remains an underlying legacy of peasant agriculture, traditional field systems and settlement forms which continues to represent an important dimension of the rural social geography of Western Europe. Although the 'final phase' of the European peasantry was announced some time ago (Franklin 1969), peasant farming still dominates many sub-regions of Mediterranean Europe, and even retains a foothold in parts of Alpine Europe, West Germany, Scandinavia and the United Kingdom. Similarly, it is still possible to find examples of the classic field systems and landholding patterns of pre-industrial Europe: Roman centuriation in the Po valley around Milan, irregular open-field systems on the Iberian peninsula, southern Italy and Greece, regular rectangular strips of *gewannflur* field systems on the loess soils of Lorraine and the Alsatian Plain, small enclosed and wooded field systems which characterise the *bocage* landscapes of Brittany and Cornwall, huge landed estates in the Scottish Highlands and their counterparts – *latifundia* – in southern Spain and southern Italy, crofting systems in the Western Isles and coastal Norway; and so on (see Christians and Claude 1979 and Houston 1963 for detailed examples). And even where these systems have been replaced or modified by nineteenth-century enclosure movements, land reform and farm amalgamation, the legacy of the old order often persists in the form of classic settlement types: the small hamlets and scattered farmsteads of mountain regions and the large nucleated estate villages associated with latifundia systems, for example.

Overlying these foundations are the newer dimensions of rural geography resulting from the differential impact and interaction of structural change in agriculture, state intervention and rural transportation. Among the resulting cleavages which are of significance from the point of view of social geography are those which contrast the remoter rural areas with the peri-urban and inter-urban areas, those which discriminate between areas affected by tourism, recreation and retirement and areas which are not, and those which differentiate between marginal, traditional agricultural regions and more prosperous and highly capitalised agricultural regions. Also of significance are the cross-cutting cleavages which differentiate between groups of households *within* particular areas: incomers/locals and transport rich/transport poor. At the intersection of these cleavages is the issue of rural deprivation, which raises the question of rural planning and state policies for rural areas. These are both examined in later sections of this chapter. First, however, attention is focused on two of the most vividly contrasting elements of rural Europe: peasants and incoming urbanites.

The Peasantry of Western Europe

There is no doubt that the European peasantry has declined dramatically in volume and economic importance. Having fulfilled a crucial role in the early stages of capitalist development (hence Marx's remark that 'agriculture is the original sin that introduced wealth to the world') and established a patchwork of vernacular landscapes and ways of life, the peasantry is now very much on the retreat, its eradication being seen as both necessary and inevitable by development theorists of all persuasions (Giner and Sevila-Guzman 1980; Mendras 1970). But, residual and outmoded as it may be, the peasantry of Western Europe has clung on to remain an important and distinctive element of European social geography. In part, the survival of peasant agriculture and peasant society is in fact a reflection of the unevenness of regional development processes. At the same time, the peasantry has survived because it has adapted to the changed economic and social circumstances of the postwar period. This, of course, begs the question of just what is meant by 'peasant agriculture' and 'peasant society' (Mintz 1973; Shanin 1971). At heart, the peasant system centres on the family and the family farm as the basic units of social and economic life. As Franklin puts it (1969: 2), 'Throughout all aspects of peasant life those things which belong to the kinship order – family, status, marriage and death – are inextricably mixed with those that belong to the economic order: occupation, inheritance and property'. Beyond this, it is difficult to generalise with any confidence because of the variety of regional forms of peasant organisation and because of their constant evolution. As a result, it is also difficult to be precise about numbers. Estimates vary wildly, with Ilbery (1981) suggesting as many as 25 million to 30 million. By any reckoning, however, there are at least 8 or 10 million people in Western Europe who are dependent on some form of peasant agriculture, most of them localised in Greece, southern Italy, the Iberian peninsula, south-central France, and the uplands and mountain areas of Austria, Switzerland and West Germany.

Following Franklin (1969, 1971), it is possible to identify three different groups common to all of these areas. First are those with marginal holdings because of land fragmentation and/or a poor environment. This is a residual group, dominated by an elderly cohort who remain on the farm through sheer inertia:

> Many marginal farms are worked by old couples, or widows and widowers, eking out their last days. For them the farm offers the shelter of a home and the comfort of the security associated with ownership of property. Collectively they constitute a social problem of major proportions. Geographically their incidence is so great in some areas that they contribute towards the creation of problem regions (1971: 22).

With a large proportion of the population having migrated to towns and cities, these residual peasants seem to have a future only as park-keepers of a cultural heritage, preserved from extinction by state welfare and pension programmes.

A second group consists of those with larger or more productive units who have been able to maintain a full-time enterprise for themselves and their family. Their success has been achieved by shedding many of the traditional characteristics of peasant organisation and culture, taking advantage of the guaranteed prices and subsidies created by state intervention in agriculture, participating in co-operative marketing schemes, sharing labour and equipment with neighbours. By avoiding direct competition with larger commercial farms, these *paysans évolués* have been able to survive in the interstices of the modern agricultural system. Their position, however, is transitional and vulnerable. Most surviving full-time peasant enterprises have to cope with the attenuation of kinship networks as some of the younger members are claimed by outmigration, while integration with the commercial economy means a reorganisation of family functions and an end to self-sufficiency (Netting 1981; Pérez-Díaz 1976). Meanwhile, economic survival means keeping the remaining family labour force permanently at full stretch, thus marking an important departure from the traditional peasant way of life in which economic crises could usually be averted through a redoubling of the family's work effort. The economic rewards are thin, with the increasing viability threshold in agriculture always syphoning off any petty capital which may be accumulated. As Müller (1964) has shown, however, it is not material rewards which motivate most peasants so much as the retention of their independence (in terms of lack of supervision) and the perceived satisfactions of a family-oriented lifestyle.

The third group consists of 'five o'clock farmers': peasants who have acquired a second source of income. Although accurate statistics are hard to come by, it is generally acknowledged that this group, unlike the others, has expanded significantly since 1945. By the mid-1960s, part-time peasant holdings in West Germany outnumbered full-time peasant farms by a ratio of 2:1 (Franklin 1971) although elsewhere the ratio was closer to 1:2 or even 1:4 (Clout 1972). Part of this group consists of peasants who have acquired a second enterprise – an inn, a transport or construction business – or occupation: post office official, Burgermeister. The greater proportion of the part-time group, however, are worker-peasants whose second source of income comes from a factory job. This arrangement was greatly facilitated by the postwar transport revolution, which not only helped to resuscitate rural craft and manufacturing industry, but also made possible daily commuter trips of 25 km or more in each direction. At the same time, certain features of the peasant family system – loyalty, obligation and cheap, disciplined labour – were conducive to the development of rural industry. At first, the worker-peasant phenomenon seemed to be an ideal solution to several problems. It (i) provided extra income for peasant families; (ii) provided a modest source of capital for their farms; (iii) helped to maintain rural population and a reasonable range of local services; (iv) helped to ease the shortage of industrial labour without adding to the congestion and housing problems of cities; and (v) generally looked like serving as a medium 'by which the transition from a peasant to a capitalist society [could be] performed without a wholesale de-rooting of the rural population or a disruption of the rural community' (Franklin 1969: 22). It also seemed to offer a natural cushion during times of industrial recession, when worker-peasants might revert, temporarily, to full-time farming. In practice, however, these advantages have had to be set against a number of problems. Land left fallow because worker-peasants did not have time to tend it – *Sozialbrache* – represented a significant wasted resource, and in some cases went on to become a haven for weeds and pests (Künnecke 1974; Wild 1983a). Some worker-peasants were tempted to over-mechanise their small plots with the funds derived from their factory jobs; and the whole phenomenon began to hinder the processes of land consolidation and farm amalgamation, thus retarding the overall economic perform-ance of the agricultural sector (Clout 1972). More important, the overall shift in the economic climate in the early 1970s altered the whole position of the worker-peasant. No longer was there any shortage of labour for manufacturing industry. On the contrary, many of the industries which had been instrumental in fostering the worker-peasant phenomenon began to shed labour as mechanisation and automation became more widespread (Frank 1983). Return to full-time farming has not been an attractive proposition for many. As worker-peasants became accustomed to the levels of living associated with industrial wage-rates, so the viability threshold of farms was steadily increasing. Faced with a choice between working harder than ever for less than before, it is not surprising that increasing numbers have abandoned farm-ing altogether, leaving their land under semi-permanent Sozialebrache in the hope that it might in future be profitable, perhaps as the site of a second home.

Meanwhile, the remaining peasantry – both full-time and part-time – must cope with the problems of social as well as economic survival. Most peasant communities have been deeply penetrated by the values and norms of urban/industrial society. No longer are the security and independence of rigid family organisation always a sufficient reward for the forfeit of personal freedom; nor is the economic cushion of the extended family network sufficient compensation for abstaining from the lifestyles and opportunities of 'modern' society. The result, of course, is a steady stream of outmigration, especially of young adults. For want of sufficient research we lack the details and miss the regional nuances, but it is clear that there is a great deal of variation in the way

— and the rate — that different peasant communities react to their predicament. In the Basque country of Spain, for example, different attitudes to the value of traditional ways of life based on the *baserria* system of inherited landholdings result in quite different outcomes. In some villages, the moral and cultural desirability of the baserria way of life has persisted, and new generations continue to be socialised into the system, despite a general awareness of the futility of peasant agriculture. Elsewhere, there is competition to escape heirship of the family baserria, leading to very localised streams of outmigration (Douglas 1971).

Traditional Settlement Forms

Like the peasantry itself, the traditional settlement forms of rural Europe have been substantially overwritten by the postwar rationalisation of agriculture and the penetration of the countryside by urbanism. They have by no means been entirely eclipsed, however, so that, like the peasantry, they represent a distinctive element of European social geography. Intensive study of European settlement has generated a variety of elaborate typologies. Most follow the classic approach of Meitzen (1895) and Demangeon (1927) in recognising a basic differentiation within Western Europe between, on the one hand, dispersed settlement types and, on the other, various forms of nucleated village. Meitzen's original thesis was that this division was the product of ethnocultural differences, with Celtic occupation being responsible for the *Einzelhof* or isolated dwelling and Germanic occupation resulting in the *Haufendorf*, or nucleated village. A third ethnic group, the Slavs, was associated with the *Strassendorf* (street village) and *Rundling* (round village). This was soon discredited as an oversimplification, and it is now recognised that the peasant village and other traditional settlement forms are the product of the interaction of several factors. Sometimes the physical environment is dominant in producing distinctive settlement forms, as in the *Terpen* of northern Germany, the Netherlands and the English Fens: small villages clustered on artificial mounds above flood level. *Marschhufendorf*, on the other hand, are long linear villages which stretch along the dykes enclosing reclaimed land in the Netherlands and the Frisian coast; while a more widespread example is the linear village situated in a narrow-valley floor.

For the most part, however, the determinants of settlement form have been socio-economic. Agricultural organisation has long been recognised as a key influence. Lefèvre (1926), for example, noted the association between the dispersed settlement in Herve (Belgium) and the region's long history of cattle-rearing; while Demangeon's (1927) first attempt to classify French rural settlement was based on the influence of different field systems. The marked nucleation of settlement under the open-field system, especially the three-field system, has in fact been stressed by many writers (Green, Haselgrove and Spriggs 1978). Forms of tenure can also be important, as illustrated by the contrast between the dispersed settlement of Tuscany (associated with the individual family units of the *mezzadria* sharecropping system) and the large, agglomerated estate villages of the south (associated with the latifundia system). The colonisation of newly occupied areas also tends to produce characteristic forms (Buchanan 1979), while the penetration of the great forested belts of Europe is associated with a particular form of linear village – *Waldhufendorf* – laid out along streams with peasant holdings stretching back into the forest. Finally, the need for defence has played an important role in conditioning the form of settlement. Fortified village sites are still apparent throughout Western Europe, from the *castletouns* of Scotland to the walled hilltop villages of Appenine Italy, Corsica and Sardinia.

As Bunce (1982) points out, this variety of forms is the product of the evolving relationship between peasant populations, local resources and local institutional milieux. He also points out that a number of scholars have suggested that settlement morphology is an important manifestation of 'community', with more nucleated forms making for – and reflecting – a strong sense of social cohesion, especially where they are focused on a central village green, square or piazza. The more detailed aspects of settlement morphology and community structure, however, are strongly influenced by regional cultural variations which elude generalisation. There is widespread agreement, however, as to the broad pattern of traditional settlement types in rural Europe (Figure 4.2). Dispersed settlement, in the form of hamlets and scattered farms, is characteristic throughout the great mountain ranges, where environmental conditions dictate low population densities. It is also the underlying form of settlement in a variety of other areas – Wales, Kent, Normandy, Brittany, the Low Countries and Lower Saxony, for example – where the initial form of colonisation (often Celtic, as Meitzen pointed out) has been maintained by subsequent cultural groups with different agricultural systems. Nucleated settlement types occur in two major belts, associated with quite different sets of factors. Most clear cut are the irregular nucleated villages of the loessic soils which extend from northern France to West Germany (and, indeed, eastwards through southern Poland to Hungary and the Ukraine), representing a general form of settlement which has been appropriate to successive systems of agrarian practice ever since the first colonisation by Neolithic farmers (Houston 1963). The second distinctive belt of nucleated settlement is focused on the shores of the Mediterranean, where defensive factors and hydrological conditions have both made for the agglomeration of rural settlement, producing very large villages in parts of Greece, southern Italy and Castile (Spain). In between these two broad belts are isolated pockets of more specific nucleated forms: Marschhufendorfer and Terpen, Strassendorfer and Waldhufendorfer.

Figure 4.2: The Traditional Settlement Forms of Western Europe

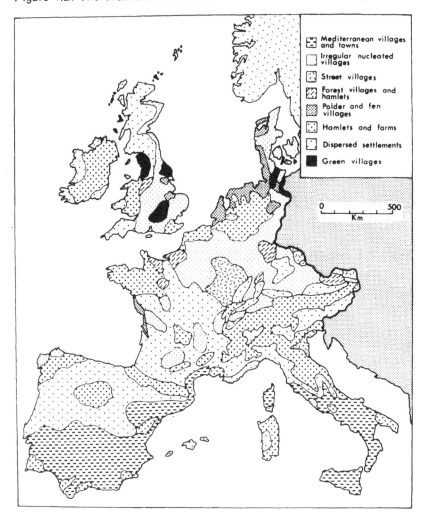

Incomers and Rural Society

Rural communities, in Western Europe as in the rest of the developed world, have traditionally been occupationally based social systems in which the family is the dominant social unit. Kinship ties have long been interpreted as the major controlling force in community life, constraining economic relationships as well as social behaviour, and creating a stable (if not static) social world, closely bound to the locality (Arensberg and Kimball 1940; Rees 1950). Social stratification has been based on property ownership, typically resulting in a hier-archical division between the landed gentry, capitalist farmers, peasant farmers and farm labourers. Some writers (e.g. Williams 1956) have suggested that these cleavages produce distinct social systems within the rural community, just as the rather different socio-economic cleavages of urban areas are associated with compartmentalised social worlds. Others, following the traditional *gemeinschaftlich* approach of rural social theory, interpret the rural community as a more closely inte-grated social system. Indeed, the whole question of rural social organ-isation remains something of a mystery, partly because community studies have become unfashionable among European rural sociologists (Newby 1980; though see Durand-Drouhin and Szwengrub 1981, 1982) and partly because of a long-standing aversion to theory in both rural geography and rural sociology (Cloke 1980; Newby 1978; Sewell 1950). Similarly, the power structure of rural areas – which determines who gets what, when and where – remains obscure; there is certainly no real evidence on which to gauge any regional variations in rural power structures. There is a general consensus, however, that the land-owning classes in each area operate an elaborate web of paternalistic relationships in order to confer stability on communities which are otherwise highly inegalitarian. It was the effectiveness of this pater-nalism which led Marx to despair of the revolutionary potential of the rural proletariat and prompted his famous remarks about the 'idiocy' of rural life and the similarity between peasants and sacks of potatoes. In the modern setting, paternalism often has to be mediated through the formal channels of local politics, as Newby, Bell, Rose and Saunders (1978) have illustrated in relation to the eastern lowlands of England. Their review showed that local outcomes, in terms of 'who gets what, when and where', were heavily biased in favour of agricultural interests in general and landowners in particular. This should not be surprising in view of the domination of landed interests in the formal political structures of the region. What is surprising, however, is that their lop-sided interpretation of the 'public interest' is rarely challenged. Newby *et al.* concluded that this is because the rural working class has remained deferential and politically fatalistic, because it is ideologically manipulated, and because those in power actively pursue a strategy of conflict avoidance. In short, the paternalism of the landowning elite has become institutionalised. Conflict can and does occur within the community, but it is mostly related to issues generated by outside intervention – the development of motorways or power stations, for example – or to 'peripheral' issues which are tolerated as a kind of safety valve for public debate: disputes over water fluoridation and the question of whether or not to open council meetings with prayers, for example.

The 'Quiet Revolution' and 'Rural Utopia'. It is into this kind of socio-political environment – occupationally based, family-centred, hierarchically stratified and, above all, stable – that the adventitious population of ex-urbanites has established itself over the past 50 years or so. As the motor car has permeated down the urban social scale, so increasing numbers of households have quit the increasingly expensive and problem-ridden cities in pursuit of cheaper housing and, above all, more secure and salubrious environments. The whole process has been reinforced by the long-standing and deep-rooted cultural Romanticism associated with rural life in Western thought (Blackwood and Carpenter 1978; Williams 1973). While cities are believed to generate chaotic social dislocation and disorganisation, rural life is believed to be characterised by social harmony, meaningful social intimacy, healthy environments and elevating surroundings. As Newby points out (1979: 23), ideas about the countryside as a visual phenomenon – the calendar/picture postcard/chocolate box stereotype – and ideas about the countryside as a social phenomenon have merged together: 'A locality which looks right must also, it is assumed, support a desirable way of life'.

The steady replacement of the indigenous rural population (displaced by structural change in agriculture) with middle-class professional and managerial commuter households and retirees has brought about a 'quiet revolution' (Ambrose 1974) in the social composition and power structure of many of the less remote rural areas of Western Europe. The countryside has been transformed – or is in the process of being transformed – into what has variously been described as a 'dispersed city', 'subtopia', 'urbs in rure'; a peri-urban environment with new social cleavages, new ways of life and new values (see, for example, Wild 1983b). A principal feature of these communities is the resulting differentiation between the social world of the established rural order and the social world of the incomers. This is clearly reflected in the spatial structure of many villages, with tracts of new owner-occupier housing forming a distinctive addition to the old village core. In Britain there is often a third component which consists of local authority housing estates (Connell 1974; Lewis 1979; Masser and Stroud 1965). As Pahl's (1965) classic work on life in a 'metropolitan village' in Hertfordshire has shown, however, it is simplistic to think in terms of a social structure which reflects the morphological divisions between established and incomer groups. He, for example, recognised six social groups:

(1) traditional landowners;
(2) local working-class households;
(3) professional and business households attracted by the idea of a rural way of life;
(4) retirees, often former urbanites;
(5) commuters who have been forced to seek rural residence because of the high costs of urban housing;
(6) working-class commuters living in older housing or on local authority housing estates.

It is fair to suggest, however, that it is the tensions between the established population and incomers which represent the most important single dimension of social life in such communities. These tensions derive as much from differing ideologies and lifestyles as income or class. Unlike the established groups, incomers do not make the village the focus of their social activities. Car ownership enables them to maintain contact with their friends elsewhere and, if necessary, to make use of urban amenities: their entertainment, their socialising and their shopping tend to take place beyond the village. Moreover, these newcomers have not entered village life as lone households which have to win social recognition from the locals in order to make life tolerable. On the contrary, their numbers and outside contacts ensure that they have little need to observe the niceties of village life. Indeed, their pursuit of urban-derived standards of consumption and amenity often leads them to 'capture' and transform some of the central institutions of established village life, from voluntary associations to school systems and local councils (Green 1964; Thorns 1968). These differences are compounded by different yardsticks of social status. The criterion by which rural workers could once enhance their social standing – skill at work – is simply not recognised by the incomers. Even *cognoscenti* of rural ways of life are rarely able to distinguish skilfully ploughed fields from indifferently prepared land; nor are they much impressed by crop yields or judgements concerning the quality of livestock. On the other hand, most of the established households cannot compete in terms of the incomers' main criterion of status – materialism – mainly because of income differentials, but also because they are less finely tuned to the rapidly changing rules of conspicuous consumption. As a result, the core of the established community turns in on itself, cutting itself off from the separate world of the newcomers by inventing a new criterion of status: length of residence (Emerson and Crompton 1968; Radford 1970).

Economic and social transactions between the two sets of groups are limited to a small knot of 'go-betweens' – shopkeepers, tradesmen and domestic workers – except where serious conflict between them emerges. Since most incomers are retired or are employed in nearby towns or cities, and since their children tend to have better educational qualifications and to be more mobile, little of this conflict is associated with the labour market. It is the housing market which is the principal arena for conflict. The supply of rural housing and building land is, typically, very limited, so that the intervention of newcomers at any level in the housing market tends to displace indigenous households and inflate prices, creating a kind of rural gentrification (Berger, Fruit, Plet and Robie 1980). In these circumstances, established low-income households become more dependent on 'tied' accommodation and, where it is available, public housing. Yet middle-class incomers often oppose the expansion of the local housing stock (especially if proposals involve public housing) on the grounds that it would be 'detrimental to the character of the rural environment' (Rodgers 1983). This points to a second major area of conflict to which reference has already been made: environmental and ecological issues. As we have seen, such conflict usually involves clashes between incomers and the farming community over the changes wrought by modern agricultural methods; except where issues of access to the countryside are concerned, when the marauding tourist or weekend motorist may be perceived as a common enemy. Ironically, the incomers themselves have often been responsible for eroding one of the attributes of rural environments which they themselves value most: the communality of village life. Having arrived with a stereotype 'village in the mind' (Pahl 1965) according to which they expect the local population to conform, incomers, by their very presence, ensure that their quest for a stable and harmonious social milieu is doomed to failure.

It must be acknowledged, however, that despite the tensions and conflicts arising from the presence of incomers, there are several major ways in which rural communities have unequivocally benefited from the 'quiet revolution'. One outcome which has already been mentioned is the generation of employment, especially service employment. The high spending power of the incomers also makes for *qualitative* improvements in rural services — helping to expand the range of foodstuffs and clothing in local shops, for example. Their affluence also strengthens the local tax base (though they may not exert a particularly progressive influence on the nature of subsequent public expenditure: see, for example, Limouzin (1975)). Finally, their numbers alone have gone a long way towards providing support for local schools, community centres and associations which might otherwise have closed in the face of rural depopulation (Emerson and Crompton 1968; Lewis 1967).

Rural Retreats: Second Homes and Alternative Lifestyles. Such benefits are hardly applicable to the impact of incomers in remoter rural areas, however. Here, beyond the commuting threshold of major towns and cities, traditional rural communities have been penetrated by relatively small numbers of adventitious households of three very different types: retirees, owners of second homes, and those in search of an alternative to ways of life based on urbanism. The movement of retirees from cities has already been mentioned (see p. 36) and at this point it is necessary only to reiterate that the bulk of migrant retirees head for established 'places of reward and repose' while many of those who do make for *remoter* rural regions turn out to be returning to their place of origin and so cannot strictly be considered as incomers.

Second homes are important only in parts of Western Europe. The phenomenon is most pronounced in France, where between 15 and 20 per cent of all households have some kind of second home, and in Scandinavia, where the proportion ranges between 10 and 15 per cent. Elsewhere, although second-home ownership is rapidly expanding, the proportion is less than 5 per cent. Largely a product of postwar affluence, second-home ownership is overwhelmingly dominated by the big-city middle classes for whom the proposition of a summer home or weekend cottage is enhanced by its investment potential and its possible use on retirement (Clout 1969). Localised in the more attractive rural regions, second homes create a kind of 'seasonal suburbanisation' (Clout 1972), introducing substantial numbers of incomers to rural communities on a temporary and intermittent basis. A good deal of this activity merely serves to reinforce the effects of the commuting incomers discussed above, since there is a marked distance-decay effect on the selection of second homes. There is some evidence, however, that the expansion of the adventitious population itself has tended to displace second home development further afield (Boyer 1980). Distance-decay effects notwithstanding, remoter rural areas are still left with a substantial number of second homes (Aldskogius 1978; Davies and O'Farrell 1981; Department of the Environment 1975). In Norway about a third of the total are located in mountain areas; and in rural areas as a whole second homes now outnumber year-round dwellings inhabited by the agricultural population by nearly 2 to 1 (Langdalen 1980).

The benefits of second homes in such areas include a modest, but seasonal, contribution to the local economy (principally in the form of trade for local shopkeepers). Clout (1972) notes that the increase in social life associated with the presence of second-home owners may also be seen as an advantage by the local community, although much will clearly depend on the particular social and cultural context. In Wales, for example, the second homes of English families have become the target for fire-raising in the cause of Welsh nationalism. This somewhat extreme example emphasises once again the more general problem of social disruption associated with incomers of any kind. Langdalen (1980) suggests that second-home development realigns local communities according to particular interest groups whose objectives may conflict, with the issues sometimes spilling over to involve outside interest groups (developers, planners, amenity groups, etc.). But the biggest single source of conflict is again the housing market. Although purpose-built second homes for rent or for 'timesharing' are becoming more commonplace, most second homes are older properties – cabins, farmhouses and cottages – which have been bought outright. Some of this housing, at least, represents an erosion of the stock available to the local population, creating price inflation which may effectively put some of the remaining stock out of reach (de Vane 1975). Langdalen (1980) notes that the conversion of dwellings to second homes also tends to distract the local workforce from ordinary housebuilding and maintenance. In short, the benefits and disbenefits of second homes to rural communities should not be regarded as compensatory, since they affect different sections of the community. The net result is likely to be a tendency towards a polarisation of well-being, with property owners, tradesmen, shopkeepers and restaurateurs deriving the bulk of the advantages; though it should be acknowledged that the evidence is fragmentary and the arguments both for and against second homes have been somewhat overplayed (Coppock 1977; Mahon and Constable 1973; Sacchi de Angelis 1979).

Finally, mention should be made of the incomers whom Chevalier (1981) has dubbed 'neo-rurals': neo-artisans and neo-peasants, deserters from the urban social order who have deliberately sought out remoter areas in an attempt to pursue 'alternative' lifestyles based on independence and self-sufficiency. As Koch (1979) puts it:

Economic crises, weakened employment prospects, increased political barriers and the failure of extensive reforms ... all this makes rural life seem once again more attractive than life in the metropolis — especially in the consciousness of students and intellectuals. Together with fantasies about the countryside, frequently romantic and unrealistic ... these factors intensify ... the search for elevating and virgin living conditions. (p. 215)

It is this group which has often been cited as the vanguard of the 'rural renaissance' and 'rural turnaround' of the past decade (Mendras 1979). The number of households involved is difficult to estimate, but it is clear that 'neo-rural' households now represent a significant proportion in some of the more isolated rural regions. In Dyfed (Wales), for instance, some parishes are now entirely populated by 'neo-rural' households of various kinds (Ford 1982). As with other types of incomer, however, their presence sometimes tends to be disruptive of the social environment they have so carefully sought out. Forsyth's (1980, 1982) study of incomers on a small Orkney island is illustrative. She found that, although numerically subordinate, incomers tended to dominate community affairs. In part, this was because they misinterpreted traditional, muted forms of social interaction as a reflection of unimaginativeness and inarticulacy. In part, it was because the newcomers were less interested in the traditional Orcadian way of life than in making Orcadians fit their notion of what rural life should be. As a result, incomers tended to take over the bulk of all local organisations, bringing with them the socio-political style of the urban south. As in other established/newcomer situations, this resulted not in an invigorated community life, but in the withdrawal of the local community and the creation of tensions focused on property and social comportment. Meanwhile, the incomers' efforts at sustaining idealised lifestyles were often less successful than had been hoped. Aspiring neo-ruralists may be able to learn about woodburning stoves, wholewheat bread and corn dollies from books and catalogues, but extracting a living from marginal environments requires experience which cannot be substituted by research or hard work. For many neo-ruralists, therefore, the ideal of self-sufficiency is compromised, with social security benefits having to bridge the gap between household production and an acceptable level of living.

Disadvantaged Rural Europe

As we have seen, structural change in agriculture has brought about a general decrease in rural employment opportunities. In many regions this fundamental problem has been intensified by the marginality of the rural resource base. Shortages of cultivable land, poor soils, difficult topography, short growing seasons and unreliable rainfall all make for economic marginality, and this, in turn, forms the basis for the classic syndrome of rural depression: low yields and uncertain production, limited enthusiasm for innovation, underemployment, low wages, high levels of long-term unemployment and, above all, *depopulation*. As Clout (1976) points out, it is very difficult to gauge with any real accuracy the true pattern of rural depopulation although it clearly represents an important component of regional change throughout Western Europe. Even relatively prosperous agricultural regions have been losing population because of the intrinsic weaknesses of the agricultural sector as a source of employment (Drudy 1978), while depressed rural regions continue to lose population at a striking rate. In Portugal, for example, the provinces of Guarda, Braganca, Vila Real and Viseu lost population at the rate of 26.8, 24.4, 18.9 and 15.4 per cent respectively between 1960 and 1970, with the total loss amounting to over 250,000 people. Similar levels of depopulation are described in the Council of Europe's *Report on Migration and Population Redistribution* (Kayser 1977) for Spain, Italy, Greece, Ireland, Finland, Norway and Sweden.

Such losses of population not only reflect people's response to the lack of opportunity in depressed rural communities — many areas, as suggested in Chapter 2, have now reached the stage where the process of depopulation itself serves to erode the quality of life of those left behind. Outmigration, always age-selective, leads to an increasingly aged, less mobile population with greater requirements in terms of medical and social services. Moreover, because of the rising threshold population required by most services and facilities (both public and commercial), the operation of many becomes marginal; and in some cases the future of a shop, a clinic or a primary school may rest on the movement of one or two families (Forsyth 1984; Knox and Cottam

1981a; Sundberg and Öström 1982). Villages with a population of between 300 and 500 can usually support a shop-cum-post office and a pub or an inn, but primary schools require a support population of around 1,000 to be economically justifiable, middle schools require a support population of 4,000, pharmacies require a catchment of 4,000-4,500, and general practitioners need about 2,000 patients in order to justify a regular surgery. In order to place these thresholds in some perspective, it is important to realise that most settlements in rural Europe contain less than 500 people. Take Norfolk, for example. Ayton (1976) calculated that 44 per cent of the villages there were below 300 in population, 66 per cent below 500, and 88 per cent below 1,000. Inevitably, there has been a withdrawal of services from some areas. Within the Highland region of Scotland (population 200,000 in 1981), for example, there was a net loss between 1971 and 1978 of 31 sub post offices, 18 primary schools and 3 secondary schools, with similar reductions in relation to medical and pharmaceutical services and to basic retail facilities such as butchers' shops and general stores (Knox and Cottam 1981a; see also Harman 1982). The consequences of such declines in service provision involve inconvenience, if not actual hardship or social disintegration. The existence of a basic set of utilities, social facilities and shops is critical to the quality of rural life, not least in terms of community cohesion; and in this context it is above all the primary school which is generally felt to embody the rural community as something alive and enduring. Without it, young families are reluctant to stay or settle, thus adding to the circular and cumulative process of decline and depopulation (Edwards 1971).

In recent years attention has shifted somewhat from the process of rural depopulation to the conditions of disadvantage and deprivation which are at once its cause and effect. Just as poverty itself was 'rediscovered' by politicians and social scientists after the postwar consumer boom, so the relative deprivation of rural areas became a salient issue as it emerged that the benefits of the rural transport revolution had been differential in their impact, and that problems associated with structural change and economic marginality had been compounded by depopulation and the increasing difficulties of maintaining rural shops and services. Rural regions, it was claimed, had 'fallen behind' the rest of Europe (Moore 1979), not only in terms of the available range of economic opportunities, but also in terms of the provision of housing, education, health and welfare facilities and the consumption of all kinds of goods and services (Council of Europe 1980). These themes have been taken up by a number of researchers, and there is now an emerging literature on various pieces of what Cloke (1980) has called the 'deprivation jigsaw' (see, for example, Knox and Cottam 1981b; Moseley 1978; Neate 1981; Shaw 1979). Ironically, the bulk of this literature is addressed to conditions in northwestern Europe, whereas it is in southern Europe that the worst extremes of rural deprivation have always been concentrated. At the same time, the emphasis on rural deprivation serves to divert attention from urban/industrial problems. Indeed, it is relevant to note that the whole issue of rural deprivation has been vigorously publicised by rural-interest groups (see, for example, Association of County Councils 1979; Standing Conference of Rural Community Councils 1978).

In fact, the distinction between rural deprivation and urban deprivation can itself be misleading. Moseley (1980) recognises this, pointing out that both are the product of fundamental economic forces and that both are characterised by similar syndromes of social problems. Both, for example, involve high levels of unemployment, low wages and restricted job opportunities stemming from local economic stagnation. These problems lead in turn to an erosion of community morale and, eventually, to depopulation, leaving behind the aged and the socially and economically less competent who become trapped in the worst of the housing stock. Meanwhile, privately organised facilities and services decline or disappear, and public sector services struggle with the dilemma of an increasingly indigent population coupled with a progressively smaller tax base. The *differences* between urban and rural deprivation stem largely from basic contrasts in the physical and social environment (Knox and Cottam 1981b). Thus, whereas the fundamental dimensions of urban deprivation are compounded by problems of environmental decay, class and ethnic conflict, overcrowding, delinquency, criminality and social disorganisation, rural deprivation is intensified by problems of inaccessibility, social isolation and the lack of a threshold population large enough to attract sufficient services and amenities. On the other hand, urban problems have to be set against the richer 'opportunity environment' of cities, while rural problems must be set against the advantages of environmental quality and social stability. Such cost-benefit analysis would be difficult to operationalise, though there is some evidence from some social surveys that rural populations tend at least to be better able to identify the compensations of their milieux (Townsend 1979; van Bemmel 1981).

Nevertheless, it is clear that poverty and deprivation can be compounded by the nature of the rural environment, particularly in *remoter*, peripheral regions. Consumer prices illustrate one aspect of this. A combination of local monopoly power, high transport costs and low turnover makes for significantly higher prices in every category of goods. Price surveys in rural Scotland, for example, have established that, compared with prices in a provincial city like Aberdeen, the price of food in rural areas tends to be about 15 per cent higher, while consumer goods are about 20 per cent more expensive. The remoter and more isolated the community, the worse the problem. Thus, in Barra (the southernmost of the Western Isles) food prices in general in 1981 were 27 per cent higher than those in Aberdeen (with alcohol and tobacco costing 33 per cent more); and consumer goods were nearly 25 per cent more expensive (Mackay and Laing 1982). In addition to high prices, remoteness and isolation also serve to compound the basic problems of poor and vulnerable households through various aspects of 'opportunity deprivation'. In relation to retailing, for instance, the limited stock carried by many rural stores means that households without freezers will enjoy a very limited choice of fresh foods. Another, more important, example of opportunity deprivation concerns education. The small schools characteristic of most rural areas may be a central focus of the local community, but they are in fact also associated with educational disadvantages in terms of socialisation and access to resources. This, in turn, can impose stress on adults, especially since compensatory facilities such as nurseries and playgroups are rarely available (Nash 1980; Sher 1978; Watkins 1979). Finally, opportunity deprivation in general is in its turn reinforced by various aspects of mobility deprivation which are endemic to remoter rural areas. Not only can transport costs loom large in family budgets because of the distances between workplace, amenities and the home; rural transport is often poorly developed, so that households without their own means of transportation can be seriously disadvantaged in relation to everything from medical care to legal advice (Banister 1980; Moseley 1979; Tykkaläinen 1981).

Rural Areas and Public Policy

State policies relating to agriculture and the rural environment have become increasingly important throughout Western Europe since 1945. In part, of course, this is a reflection of the general trend towards the increasing role of the state in most spheres of life. There are, however, more specific reasons which have been given for state intervention in rural affairs. The most common of these stem from the fundamental economic objectives of increased efficiency in agriculture coupled with the 'management' of production in order to avoid gluts and shortages.

The casualties of the structural transition in agriculture represent a second major reason for state intervention. On the one hand there have been the immediate casualties: small-scale operators who have become 'marginalised' by the rising viability threshold of farm units. On the other there have been the indirect casualties: entire communities which have suffered from the effects of unemployment and depopulation. More recently, recognition of the inherent disadvantages of peripherality has provided a further reason for intervention as the issue of rural deprivation has become salient. At face value, then, most public policy can be interpreted in terms of attempts at economic management and/ or as expressions of social concern (Wallace 1981). It is important to bear in mind, however, that state intervention can also be interpreted as a means of fostering capital accumulation and legitimising the overall structure of the political economy. In relation to agricultural policies, for example, state intervention on behalf of small-scale operators can be seen as an attempt to maintain 'the mirage of the viability of small business'; while intervention related to the 'efficiency' and 'management' of agriculture can be seen as attempts to encourage capital accumulation not only within the agricultural sector but also within the broader economy (Buttel 1980).

Whatever the interpretation, it is clear that state intervention has played an increasing role in recasting and redirecting much of the social geography of Western Europe. In practice, there has been a complex variety of forms of intervention, with different governments pursuing different combinations of policy measures according to their own economic circumstances and social values (Bowler 1979). As a result, it is virtually impossible to evaluate their impact on different regions or social groups. It is appropriate, however, briefly to illustrate the nature and scope of the more common components of state intervention.

Land Consolidation and Farm Enlargement. One of the earliest forms of state intervention in rural Europe, land consolidation programmes have now been undertaken in every country in an attempt to eradicate the inefficient fragmentation of individual holdings resulting from partible inheritance laws and the inertia of outmoded farming systems. Finland and Denmark were the first to legislate for consolidated holdings in place of fragmented open-field systems, and they were soon followed by the Enclosure Movement of nineteenth-century Britain and by parallel legislation in northern Germany, Switzerland and Austria. The most extensive land-consolidation programmes this century have been those of France, Spain and West Germany, although the programmes vary a good deal in their scope and approach from simple exchanges of parcels of land to more ambitious schemes in which land consolidation is just one of an integrated series of agricultural reforms. In France, an official policy of land consolidation – *remembrement* – was launched in 1941, when it was estimated that 40 per cent of all farmland was in need of reorganisation. By 1970, 43 per cent of this (over 6 million ha) had successfully been consolidated. Further progress has been slow, however, largely because of the conservatism of land-owners. Similarly, consolidation programmes in Spain and West Germany involving 10 million ha and 9 million ha respectively are expected to take at least until the end of this century to complete, leaving some regions with a highly inefficient system of spatial organisation in other-wise favourable agricultural environments. A recent study of Galicia, for example, showed that each farm, on average, is fragmented into over 30 separate parcels of land. On the other hand, it must be acknow-ledged that land fragmentation does at least retard the process of depopulation. Moreover, there are some environments – such as the Alpine region and the more peripheral regions of the Mediterranean – where fragmentation remains closely tuned to local needs.

The processes behind land fragmentation have also been responsible for the prevalence of undersized farm units in much of Western Europe. Land consolidation goes some way towards resolving this problem since it paves the way for farm amalgamation. In some countries, however (Austria, France, the Netherlands, Sweden and West Germany), addi-tional programmes have been introduced in order to facilitate farm enlargement. In some cases, legislation ensures that when farmland falls vacant it is used to enlarge neighbouring holdings. In others, pensions or retraining grants are offered to farmers who sell up their land for amalgamation; while in France and Sweden there are public agencies which systematically buy up land in order to create land banks which can later be used to facilitate farm enlargement (Bergmann 1970; Bowler 1983; Hirsch and Maunder 1978; King and Burton 1982, 1983; O'Flanagan 1980).

Tenure Reforms. The reform of the social (as opposed to the spatial) organisation of land ownership has been a central objective of state intervention in Italy, where attempts have been made to dismantle the massive, under-utilised latifundia in order to provide farms for landless agricultural labourers and to enlarge neighbouring *minifundia*. The objective has been to reinforce rural democracy and raise levels of rural social well-being while at the same time rationalising the agricultural base. Tenure reform was a central component of a complex package of public policies set up in 1950, which also included land reclamation, land settlement and infrastructural schemes. This legislation encompassed over 8 million ha, nearly 30 per cent of the total land surface. In practice, however, only 1.2 million ha of land has been affected by the policies, and of this less than 770,000 ha have been appropriated and distributed to a total of 113,066 families. The land resources thus obtained have been insufficient to promote a radical improvement in either agricultural efficiency or rural social well-being, however. Moreover, many of the newly created farm units have turned out to be expensive to acquire, and/or inadequate in size to employ a family and provide sufficient income. As a result, state policy has been redirected towards the creation of employment in manufacturing industry in rural regions (Diem 1963; King 1971, 1973; Maos 1981).

Land Settlement and Infrastructural Improvements. These have been important in Mediterranean Europe, where a large 'surplus' of rural population is coupled with extensive tracts of unimproved land. Spain, in particular, has relied heavily on integrated projects involving hydro-electric schemes, river control, irrigation, afforestation, roadbuilding and the resettlement of peasant farmers and farm labourers in purpose-built farms on improved land as a solution to the problems of the retarded agricultural regions of the south and southwest. But although Spain's hydrological schemes are technically of the highest order, they have proved expensive and relatively ineffective in ameliorating the socio-economic problems of disadvantaged rural regions. As in other countries, such schemes have been unable to draw sufficient amounts of industry or commerce in their wake, while newly settled farmers have been given insufficient support in terms of marketing networks and pricing policies. In addition, many of the settlers themselves were unprepared for the demands of their new lifestyle. This has led, in some cases, to the abandonment of the new, purpose-built but isolated farm-houses as farmers have returned to their agro-towns as a base from which to work the land (Aceves 1976; McEntire and Agostini 1970; Richardson 1975).

Settlement Rationalisation. Attempts to respond to the increasing viability threshold of services and amenities through the planned rationalisation of rural settlement have been pursued to some degree by most countries. The major thrust of public policy in this context has been the designation of 'trigger areas' and 'key settlements', where investment in employment, infrastructural improvements and service provision is concentrated in the hope of creating rural growth points or, at least, retaining an adequate range of employment and retailing opportunities. Such policies have been pursued most vigorously in the United Kingdom, where a long-standing concern for rural-settlement planning was given considerable impetus by the Town and Country Planning Act of 1947. But here, as elsewhere, there has been increasing opposition to settlement concentration, partly because the corollary of support for some areas has been the withdrawal of support, implicitly or explicitly, for others. This was epitomised by the planned liquidation of 'non-viable' villages in County Durham during the 1950s and 1960s: the so-called Category D villages. Further criticism has been directed towards the inadequacy of available techniques of policy implementation and evaluation, and there is now a growing body of planning opinion in favour of settlement *dispersal*, reflecting the recent emphasis on problems of accessibility and rural deprivation. Indeed, several countries – including Denmark, Switzerland and West Germany – have already initiated 'maximum accessibility' approaches as an alternative to the planned concentration of rural settlement, while others have begun to give greater emphasis to the improvement of rural transport services as a means of improving accessibility (Banister 1983; Cloke 1979; Cresswell 1978; Flatrès 1974; Woodruffe 1976).

Rural Development Agencies. In some cases national governments, recognising the intensity and complexity of social and economic problems in particular regions, have created semi-autonomous agencies with responsibility for formulating and co-ordinating strategic policies. The best-known example is undoubtedly the Italian *Cassa per il Mezzogiorno*, set up in 1950 to encourage development in the South by complementing the normal activities of the state through so-called 'extraordinary interventions' whenever the standard state apparatus was felt to be inadequate or inappropriate. In its initial phase the Cassa concentrated its resources on improving the rural infrastructure and on backing up the tenure-reform programme (see above). Since 1957, however, its efforts have been redirected towards attracting tourism and manufacturing industry to the rural South through a combination of capital grants, tax concessions and soft loans, together with legislation which compelled the Italian state-holding sector (including the likes of Alitalia, Alfa Romeo and the Bank of Rome) to have at least 40 per cent of its total investment in the South and to place at least 80 per cent of its new investment there.

Other well-known examples of rural development agencies include the Fund for North Norway and the Highlands and Islands Development Board for Scotland (HIDB). Like the Cassa, both have steadily grown, both in terms of the size of their budget and the scope of their activities. All three, however, also share in having been unable to effect the relative improvements in rural social well-being which were anticipated at the time of their creation. In part, this reflects the magnitude of the problems with which they have been confronted. In part, however, it is because they have been unable, or unwilling, to undertake policies which are sufficiently radical. The HIDB provides a good example in this respect. Although it was set up with powers of compulsory purchase in recognition of the inefficient and inequitable organisation of land in the Highlands, it has never used them, much to the relief of the large landed estates. Instead, it has concentrated on a combination of growth-pole strategies, the promotion of tourism and rural crafts, and, more recently, the establishment of community retailing co-operatives. Meanwhile, there are still landlords who prevent tenants from clearing scrub and improving the soil on their grazings for fear they upset the grouse; who knock off the roofs of empty tied cottages rather than pay property taxes or let them to tenants who may be a threat to the game; and who represent a major obstacle to agricultural development in the Highlands (Bird 1982; Bryden and Houston 1976; Chapman 1976; Geddes 1979; Wade 1980).

Economic Policies. Attempts to manage levels of demand and supply in the agricultural sector are arguably the most important of all aspects of state intervention in terms of their socio-economic implications for rural areas. Every national government employs some combination of price-support mechanisms, production subsidies, production quotas, import quotas, marketing quotas and the like, in an attempt to satisfy a series of rather conflicting imperatives. Chief among these are the desire to increase self-sufficiency in agriculture, to increase agricultural productivity, and to ensure an adequate livelihood for the farming community. Within the European Community such policies have been harmonised through the mechanisms of the Common Agricultural Policy (CAP), the major component of Community economic policy and the recipient of over 70 per cent of the Community's total budget in 1980 (Table 3.1).

Although the CAP has been characterised by the Community itself as 'a system of support of farmers' incomes mainly through the support of market prices' (Commission of the European Communities 1980a: 3), it has also provided the framework for attempts to encourage efficiency and self-sufficiency. In broad terms, the CAP has been successful on all three fronts. In the process it has also caused a realignment of the geography of agricultural production within the Community and a reorientation of trading patterns between Community members and the rest of the world. More important, in the present context, are the socio-economic problems generated by the operation of the CAP. The most notorious of these concerns the production of surpluses. Prices set to give a reasonable return to producers on small farms have been so favourable as to result in 'mountains' of butter, wheat, sugar and milk powder and 'lakes' of wine, to be sold off at a loss to Eastern European countries or to be 'denatured' (rendered unfit for human consumption) at considerable cost. The policy of price support has also resulted, inevitably, in transfers of income from taxpayers to producers and from consumers to producers (Morris 1980). Moreover, there is now plenty of evidence to show that these transfers of income are regressive within member countries and inequitable between them. Thus, expenditure on food generally accounts for a larger proportion of disposable income for poorer households than for the better-off. Producers, on the other hand, benefit from price-support policies in proportion to their total production, so that the larger and more prosperous farmers receive a disproportionate share of the benefits. International inequity arises because countries or regions which are major producers of price-supported produce receive the major share of the benefits while the costs of price support are shared among member nations according to the overall size of their total agricultural sector. Furthermore, the pricing system of the CAP made no concessions, until recently, to the variety of agricultural systems practised on farms of different sizes and in different regions. As a result, areas with large and/or intensive or specialised farm units (such as northern France and the Netherlands) have benefited most, together with regions specialising in the most strongly supported crops (cereals, sugar beet and dairy products). Effectively, this has meant that the richest agricultural regions have benefited most from the CAP (Cuddy 1981); not surprisingly, farm-income differentials within member countries have been maintained, if not reinforced (Table 4.1).

Not surprisingly, there has been widespread criticism of the CAP from a variety of sources. In response, some modifications have been introduced to Community agricultural policy. The *dualism* of contemporary European agriculture has been recognised, with price policies now being used to orientate the production of the bigger, more

Table 4.1: Agricultural Income Spread in Four Nations, 1970 and 1977

West Germany

Bundesländer	Income of farmer and his family per family labour unit	
	1972/3	1976/7
Schleswig-Holstein	134	128
Niedersachsen	107	121
Nordrhein-Westfalen	101	120
Hessen	82	81
Rheinland-Platz, Saarland	90	74
Baden-Württemberg	94	89
Bayern	96	86
West Germany	100	100

France

Circonscriptions d'action générale	Gross farm income per family worker	
	1970	1977
Île-de-France	388	298
Champagne-Ardennes	247	278
Picardie	247	267
Haute-Normandie	148	118
Centre	135	96
Basse-Normandie	85	65
Bourgogne	103	122
Nord	146	147
Lorraine	94	93
Alsace	69	105
Franche-Comté	87	73
Pays de la Loire	84	75
Bretagne	84	85
Poitou-Charentes	99	84
Aquitaine	65	76
Midi-Pyrénées	63	70
Limousin	50	68
Rhône-Alpes	77	76
Auvergne	62	73
Languedoc	98	150
Provence-Alpes-Roussillon; Côte d'Azur	156	124
Corse	114	236
France	100	100

Table 4.1: Contd.

United Kingdom	Net farm income per farm, excluding horticulture	
	1970/1	1976/7
England North Region ⎤		126
England East Region ⎬	108	111
England West Region ⎦		77
Wales	74	71
Scotland	110	127
Northern Ireland	71	80
United Kingdom	100	100

Italy	Gross value-added per agricultural worker	
Regioni	1971	1977
Piemonte	93	96
Valle d'Aosta	65	69
Liguria	165	113
Lombardia	157	183
Trentino-Alto Adige	71	100
Veneto	143	156
Friuli-Venezia Giulia	109	139
Emilia-Romagna	129	159
Marche	67	76
Toscana	106	103
Umbria	82	94
Lazio	128	106
Campania	89	78
Abruzzi	91	98
Molise	54	42
Puglia	75	71
Basilicata	60	52
Calabria	63	78
Sicilia	95	75
Sardegna	124	98
Italy	100	100

Source: United Kingdom Select Committee on the European Communities (1980)

prosperous units while grants, rebates and pensions have been introduced in an attempt to thin out and rationalise the smaller, marginal units. The problems of depressed rural *regions* have also been recognised with the adoption of the Less Favoured Areas Directive in 1975. This provides for an annual compensatory allowance to farmers and for grants and soft loans for farm modernisation in order to maintain farming as a viable occupation 'in areas where natural conditions are less favourable for agricultural production than elsewhere, and where farming has a fundamental role in the continued conservation of the countryside' – though Cuddy (1981) suggests that there has in fact been very little indication of an effective compensatory outcome in terms of regional Community expenditure per person employed in agriculture. In addition, the Community has recently proposed (1979) the creation of integrated rural development programmes in three areas (the Western Isles of Scotland, the department of Lozère in France, and the province of Luxembourg in southeastern Belgium) where the policy instruments of the CAP are to be combined with those of the Community's Regional and Social Funds in an attempt to promote a mixture of agricultural and non-agricultural projects, thus heralding the possibility of a truly *rural*, as opposed to simply agricultural, policy (Bowler 1976a, 1976b; Bresso 1980; Buckwell, Harvey, Thomson and Parton 1982; Rogers and Davey 1973).

5 THE INDUSTRIAL BASE AND REGIONAL SOCIO-ECONOMIC DEVELOPMENT

In terms of the world economy, Western Europe is the archetypal industrial region. The cradle of modern industrial development, it was the first to experience the transition to a 'modern' economic base, with a 'modern' demographic and occupational structure and all the hallmarks of what has come to be called 'industrial' society: a highly skilled and well-educated population with high levels of productivity and consumption, highly stratified in terms of economic status, deeply penetrated by institutional organisation, and mobilised around mass political movements. Having preserved much of its initial advantage, Western Europe as a whole remains one of the most highly industrialised and productive regions of the world. In 1980 over 25 per cent of the economically active population were engaged in manufacturing activity, compared with 22 per cent in the United States and 19 per cent in Australia. Within Western Europe, however, this figure varies considerably, from less than 20 per cent in Ireland to over 30 per cent in Switzerland and West Germany. These variations in the size of the industrial employment base, together with variations in the *nature* of the industrial base from nation to nation and region to region have been fundamental determinants not only of the pattern of regional economic development within Western Europe but also of the associated social and cultural landscapes. It is important, therefore, to trace the relationships between industrial development and regional differentiation. In this chapter the spatial impact of industrialisation is first outlined, with particular emphasis on the events of the postwar period. This is followed by an examination of the patterns and processes characterising regional economic differentiation in contemporary Europe, with a final section dealing with the public-policy response to regional problems.

Phases, Waves, Cycles and Regions

There is a popular view, propagated by many geography textbooks, of the industrial revolution as a unitary phenomenon which spread from Britain progressively to encompass more and more of Western Europe and culminating in a 'mature' stage of economic development. Regional

123

differentiation is thus seen, crudely, as a reflection of different levels of maturity in industrial development. Such a view tends to obscure the true relationships between economic change and regional development. There was not one industrial revolution but several distinctive transitional phases, each impacting to a different degree on different regions. As innovations shifted the margins of profitability in different kinds of enterprise, so the whole nature of industrial capitalism altered, with parallel changes in the pattern of class relations and, indeed, in the whole superstructure of culture and social organisation. In some cases, and for certain periods, such changes have been superimposed one on the other in the same region. In other cases, the imperatives of profitability have excluded whole regions from particular phases while exposing others to industrial development for the first time. Meanwhile, regional differentials have in turn helped to condition the overall process of economic development. What follows is a brief outline of the formative stages of industrial capitalism in Western Europe: stages which were crucial in establishing the template for regional development in the postwar period.

Europe's Industrial Transition

Sidney Pollard, in his analysis of the economic history of Western Europe, has identified three major *waves* of industrialisation, each consisting of several *phases* and each highly localised in their impact. 'Above all,' he remarks 'the industrial revolution was a regional phenomenon' (1981: 14). Indeed, the springboard for the first wave of industrialisation, which began in Britain around 1760, consisted of local concentrations of 'proto-industrialisation'. In part, this early industrial activity was localised because of the pull of mineral resources and water power. In part, it was dictated by the principle of comparative advantage, with industry being displaced into areas which had never been exploited by agriculture, or which had become disadvantaged by the agricultural changes of the previous century. This pattern of proto-industrialisation, with its external economies, infrastructural advantages and well-developed markets, helped to determine the loci of industrial development precipitated by the first phase of the first wave of industrialisation between 1760 and 1790. These included north Cornwall, eastern Shropshire, south Staffordshire, North Wales, upland Derbyshire, south Lancashire, the West Riding of Yorkshire, Tyneside, Wearside and Clydeside. As Pollard points out, although these sub-regions shared the common impetus of certain key innovations, each retained its own distinctive technological traditions and industrial style. Much of the required capital was raised locally, labour requirements were drawn (in the first instance) from the immediate area, and industrialists formed themselves into regional organisations and operated regional cartels. From the start, then, industrialisation was articulated at the regional level; and this has been a feature of subsequent phases and waves. The second phase of the first, 'British', wave, between 1790 and 1820, reinforced the position of those embryo industrial regions with a coalfield base, and saw the emergence of Ulster and South Wales as industrialised regions. Meanwhile, the prosperity of three of the early starters – North Wales, eastern Shropshire and upland Derbyshire – declined markedly as their relative advantages were eclipsed by new technologies. The third phase of the 'British' wave, between 1820 and 1850, was dominated by the expansion of the railway system. This did not foster any new industrial regions, but it did widen the market area of the existing industrial regions, at once drawing more of Britain into the sphere of industrial capitalism and further exaggerating the differences between the industrialised regions and the rest. Meanwhile, the bulk of all industrial employment – even in the core areas of the new industrialism – was still typically based on the kind of organisation and technology that had been familiar 100 years earlier, thus presaging the complex overlaying and interdigitation of economic organisation that has come to characterise the modern industrial economy.

Just as this wave of industrialisation was based on localised concentrations of proto-industrialisation, so the second wave was launched from the proto-industrial regions of 'inner Europe'. Here, the first phase of industrialisation, from around 1850, was concentrated in the Sambre-Meuse region of Belgium and in the valley of the Scheldt in Belgium and France. Subsequent phases saw the spread of industrialisation to the Aachen area, the right bank of the lower Rhine around Solingen and Remscheid, the Ruhr, Alsace, Normandy, the upper Loire valley and the Swiss industrial district between Basle and Glarus. As Pollard notes, these regions were differentiated one from another not only by their different mix of industries, but also by what he calls the 'differential of contemporaneousness', whereby the same innovations simultaneously reached local economies at very different stages of development. The result was that each innovation was received differently in different regions, with different socio-economic outcomes. In general, however, the cumulative impact of innovations in first and second-wave industrialisers made for convergence: the French Nord began to look increasingly like the central belt of the Scottish lowlands, and the Ruhr began to look increasingly like the Sambre-Meuse area. On the other hand, there was increasing divergence between those areas which had adopted an industrialised base and those which had yet to follow suit.

Some of the latter were incorporated in the third wave of industrialisation. They included parts of Britain, France, Belgium and Germany which had not been directly affected by the first two waves, together with most of the Netherlands, southern Scandinavia, northern Italy, eastern Austria and Catalonia. In these regions the spread of the railway system was particularly important in facilitating participation in the industrial transition. The path of industrialisation in these regions was also different in several other important respects. There was little by way of antecedent development on which to base industrialisation, apart from one or two small enclaves (around Vienna, Barcelona and Milan-Turin). At the same time, the relative amount of capital they were required to find in order to support industrialisation was ten times greater than during the first wave. This, together with the increasing sophistication of industrial technology and its related services, compelled the state to take on ever increasing responsibilities (Gerschenkron 1966). The economic role of the state among the later industrialisers therefore tends to be more pronounced than in the countries of 'inner Europe', though they too experienced a marked increase in state economic activity between 1870 and 1914. The residual territories of Western Europe — most of the Iberian peninsula, northern Scandinavia, southern Italy and the Balkans, which Pollard terms the 'outer periphery' — remained, like the interstices of 'inner' and intermediate Europe, mainly outside the fold of industrial capitalism, to be penetrated to different degrees over the next 50 years.

Interwoven with the spatial articulation of the expansion and metamorphosis of industrialism were the effects of regular fluctuations in the capitalist economy: the so-called Kondratieff cycles (Hall 1981a; Kondratieff 1935; Schumpeter 1939). These have been shown to have a periodicity of around 50 years from boom to recession and back to boom again, with a long climb up from each recession being followed by a sharp slide into another one. During the recession period of each cycle, it seems, there have occurred exceptional clusters of new inventions, the application of which, as clusters of innovations, have initiated succeeding Kondratieff upswings. According to Mensch (1983), key clusters of innovations occurred in the years 1764, 1825, 1866 and 1935, to be followed between 11 and 17 years later by a climb away from recession. The first cycle was associated with Abraham Day's discovery of smelting iron ore with coal and with the mechanisation of the English textile industry. The second was associated with steam engines, railways and Bessemer steel; the third with electricity, chemicals and automobiles; and the fourth with electronics and aerospace. According to Mensch, the fifth Kondratieff cycle should begin in 1989, and will probably be based on the likes of microchip technology, biotechnology, and new energy-related technologies such as heat pumps and solar-energy systems. The point about these cycles, of course, is that the ups and downs of each have been imprinted on the economic and social landscape of Western Europe, contributing a significant dimension to the framework of uneven development inherited by the postwar period of 'late' or 'post-industrial' capitalism.

Postwar Change in Europe's Economic Geography

The dominant characteristic of late capitalism has been a shift in emphasis from manufacturing activity to service activity in the tertiary and quaternary sectors. As the application of technology and capital investment in the manufacturing sector has increased levels of productivity, so the demand for labour has slackened off. On the other hand, more people have been required to advertise, handle, sell and service manufactured goods, thus expanding the tertiary sector. Meanwhile, the evolution of large business organisations, the expansion of state activities and the increasing functional and technological complexity of the economy have required a marked increase in the quaternary, or information-handling sector: finance, education, research, public administration and so on (Aldcroft 1978). Initially, these shifts were the result of faster relative growth in the tertiary and quaternary sectors compared with the slackening growth-rate of manufacturing industry, with all three gaining relatively from the continued decline in agricultural employment. In recent years, however, the proportion of the workforce employed in manufacturing in some countries has begun to fall (Figure 5.1). This has been most pronounced in northwestern Europe, particularly in the United Kingdom, where the contraction and reorganisation of the Lancashire textile industry alone has cost half a million jobs since 1945 (D. Smith 1979). Between 1966 and 1976 more than one million manufacturing jobs disappeared in *net* terms, a fall of 13.5 per cent. This decline affected almost every sector of manufacturing: not just the 'traditional' pillars of the manufacturing sector — shipbuilding (−9.7 per cent), metal manufacture (−21.3 per cent), mechanical engineering (−14.5 per cent) and textiles (−27.6 per cent) — but also its former growth sectors and the bases of the fourth

Kondratieff cycle – motor vehicles (–10.1 per cent) and electrical engineering (–10.5 per cent). The experience of the United Kingdom is certainly extreme, and may have something to do with its exceptionally low export performance in the postwar period. On the other hand, the clear general trend towards a general decline in the manufacturing sector reflected by Figure 5.1 has prompted the suggestion that the advanced economies of Western Europe have in fact entered a phase of *de-industrialisation* (Blackaby 1979; Goddard 1983).

Figure 5.1: The Decline of Manufacturing Employment, 1950-80

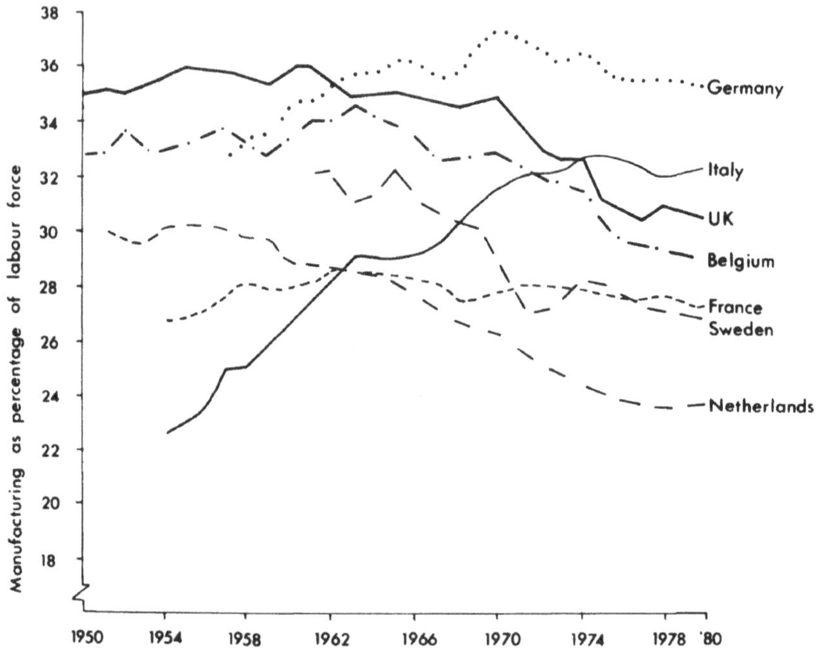

Source of data: National statistical abstracts.

What is clear is that these sectoral shifts have precipitated important changes in the character of modern industrial society (Dicken and Lloyd 1981). The *nature* of employment has changed, with relatively less people engaged, overall, in unskilled and manual jobs and more in white-collar jobs, especially professional and technical jobs. This, in turn, has had implications for several critical aspects of social well-being. White-collar jobs seem to be much less susceptible to cyclical economic fluctuations, for example, and are generally associated with less hazardous and more congenial working conditions. Moreover, the shift from unskilled to skilled work and the increasing professionalisation of all economic sectors are based upon wider access to higher standards of education. Educational opportunities, therefore, have become more important to social mobility, economic advancement and personal status. Meanwhile, the expansion of higher education has undoubtedly affected people's lifestyle preferences and consumption patterns.

The growth of the tertiary and quaternary sectors has also contributed towards drawing more women, particularly married women, into the workforce. As Dicken and Lloyd (1981) put it, this has amounted to a major social revolution

> which has affected not only the economy but also society as a whole. One of the characteristic features of modern Western society ... is the emergence of the multiworker family where both mother and father have full-time occupations and the children, according to age, find a variety of short-time, part-time and full-time money-earning occupations. A society dedicated to consumerism and the advertising media has perhaps 'hooked' a more substantial number of its members on the need to earn money for the 'good life' and, in the process, second household incomes have moved from a luxury or necessity (depending on social position) to a norm in most families. (p. 138)

Regional Patterns

The regional imprint of late capitalism has been mediated through the intervention of several key processes, including corporate reorganisation and government intervention. At the same time, the outcomes have been conditioned by the framework of uneven development inherited from earlier waves and cycles of economic development. Most regional economies, as we have seen, had become specialised to some degree during the process of industrialisation; consequently, they were vulnerable to sectoral decline, particularly in the first phases of economic readjustment and reorganisation after 1945. Similarly, regions which had become specialised in labour-intensive industries such as textiles and clothing have been vulnerable to the increasing internationalisation of capital which has rearranged the spatial division of labour on a global scale (Frobel, Heinrichs and Kreye 1980). On the other hand, the impact of the expanding tertiary and quaternary sectors has been channelled towards metropolitan locations which have been able to offer the kind of environment and workforce required by information-processing, co-ordinating and controlling activities (Daniels 1982; Marquand 1980; Penouil and Petrella 1982).

The spatial impact of corporate reorganisation stems from the emergence during the postwar period of giant manufacturing corporations whose different functional components have become geographically differentiated. Two aspects are particularly important. The first is the continuing spatial concentration of higher-order corporate functions such as headquarters administration and research and development (R and D) units. There is now a marked localisation of European corporate headquarters in London, Brussels, Paris, Hamburg and the Rhine/Ruhr area: a pattern which has become more pronounced over the past 25 years through acquisitions and mergers of companies (Daniels 1982). It should be noted, however, that there is some evidence of locational shifts occurring within these major 'control regions'. Pushed out from central-city locations by the escalating costs of office space, a number of headquarters and R and D units have been attracted to cheaper and 'amenity-rich' locations on the margins of the metropolitan area and in small and medium-sized cities in central and intermediate regions – Reading, Cambridge, Portsmouth/Southampton, Rouen, Amiens, Orléans and Tours, for example (Hall 1981b).

Conversely, the provincial regions of Western Europe tend to be dominated increasingly by the branch-plant production units of large, multi-locational corporations whose headquarters are located elsewhere (see, for example, Camagni and Cappelin 1981; Friis 1980; Hudson 1983a; Perrons 1981). The result has been the creation of what have been called 'branch-plant economies' in the peripheral regions of several countries. Hudson (1983b) distinguishes between branch-plant industrialisation proper and diffuse industrialisation. The former is based on the skilled manual-labour reserves of declining industrial regions. The latter is based on the reserves of unskilled labour in peripheral agricultural regions, where wage levels and rates of unionisation both tend to be low. Central and northeastern Italy provides a classic example of diffuse industrialisation, much of it resulting from the decentralisation of companies from northwestern Italy in response to the increasing shortage, cost and militancy of labour there in the early 1960s (Arcangeli, Borzaga and Goglio 1980; Bagnasco 1982). Typically, it involves branches of production in which labour costs are an important part of overall production costs, and in which there has been little scope for reducing labour costs through technological change: clothing, leather and wood processing, for example. The combined impact of these two aspects of industrial reorganisation has been considerable. In the United Kingdom, for example, it has been estimated that over 80 per cent of all industrial moves to the peripheral regions during the 1960s involved branch plants (Keeble 1976). Mergers and acquisitions have reinforced this trend towards the geographical separation of manufacturing activities. They have also drawn attention to the *external control* of peripheral economies which has arisen from the development of branch-plant economies. In an early analysis of this phenomenon, Pred (1974) found that in the 50 years from 1914 the production of externally controlled manufacturing jobs in Norköping (Sweden) had risen from 5 per cent to over 50 per cent. Similar trends have occurred elsewhere in Europe's peripheral regions. In Scotland, for example, over 60 per cent of the manufacturing sector is externally controlled (Firn 1975), while in the northern region of England the figure is nearer 80 per cent (I. Smith 1979).

Transnational corporations have been particularly important in influencing the extent and the spatial pattern of external control. Paine (1979) has pointed out that, while much of the locational reorganisation of transnational corporations has involved the establishment and acquisition of branch plants in the *global* periphery, there has been a significant sub-pattern of investment in the peripheral regions of Western Europe. In part, this has been a response to the termination of the migrant labour system within Europe. Other important influences *within* Europe include the structure of tariff barriers and the location of key resources such as North Sea oil and Greek bauxite (Vaitsos 1979). In detail, however, the spatial pattern of transnational activity is highly complex. Not only do the branch plants controlled by companies from different countries reflect different preferences for different countries within Europe (Hamilton 1976; Dicken 1980); they also reflect different locational preferences for peripheral locations *within* each country. Blackborn (1972), for example, found that while British-owned factories in Ireland tend to be heavily concentrated around Dublin, American-owned factories tend to be localised in government-sponsored industrial estates in development areas, and German-owned factories tend to be attracted to more peripheral locations in the small, remote towns of the southwest (see also Dicken and Lloyd 1980; Kemper and de Smidt 1980; Law 1980; O'Farrell 1980).

There is no doubt that the creation of branch plants by transnational corporations and large domestic firms has created large numbers of jobs in certain regions. It has become a moot point, however, as to whether such a development will be of long-term benefit to the regions involved.

On the positive side, it can be argued that local branch-plant economies will benefit by having access to the financial resources and technological and administrative innovations of the parent firm (Watts 1981). Conversely, it must be acknowledged that much of the expansion of branch plants has been at the expense of redundancies elsewhere — particularly in the inner-city areas of metropolitan regions — as firms have restructured their operations in attempts to divest themselves of obsolescent capital equipment and to seek out a cheaper and/or more docile workforce. As far as the local branch-plant economies themselves are concerned, it has been suggested that the absence of 'higher-order' corporate functions limits the profile of local employment opportunities, leading to a 'deskilling' process (Massey and Meegan 1982), to the suppression of entrepreneurial drive and enthusiasm (Firn 1975; McDermott 1979), and to the retardation of technological innovation (Ewers and Wettmann 1980). It has also been pointed out that a high degree of external control will result in a very open regional economy, so that international economic fluctuations are transmitted into the region relatively quickly (Firn 1975). The corollary of this is that externally controlled plants are poorly integrated with the regional economy, thus attenuating any potential multiplier effects. Others have suggested that the cumulative effect will be to depress relative levels of regional income and consumption, the knock-on effect of which may lead to a downward spiral of socio-economic disadvantage.

The other major influence on the contemporary geography of employment which should be mentioned here is that of government policy. Regional economic policies are discussed in more detail below (pp. 145-56). By encouraging investment of certain types in some areas and discouraging it in others, they have undoubtedly had a considerable impact on geographical patterns of employment. It is important to bear in mind, however, that other kinds of government policies have also had a significant impact on the geography of employment. As very substantial employers in their own right, governments clearly exert a direct influence on employment patterns, with a particular bias towards the quaternary sector in general and towards female clerical work in particular. Much of this employment is localised in national and provincial capitals, but some governments have deliberately decentralised some of their functions. The Danish government, for example, moved its regional policy administration to Silkeborg in Jutland in 1967, while the United Kingdom Government has carried out a more substantial dispersal programme, including the removal of vehicle licensing and social service administration from London to Swansea and

Newcastle upon Tyne respectively. Similarly, some governments have been able to influence patterns of employment by way of locating or relocating plants belonging to nationalised industries; by maintaining operations in plants which would be considered non-viable on strict economic grounds (Maunder 1979; Sheahan 1976); or by strategically locating major educational institutions, such as the Ruhr University in Bochum, which brought several thousand new jobs and partially compensated for the complete disappearance of the local coal industry (Thomas and Tuppen 1977). Government purchases from the private sector − aerospace and shipbuilding, for example − have also been an important influence on local employment opportunities; and even policies on taxation and interest-rates have a differential effect on the economy which is eventually translated, because of the regional division of labour, into a spatial impact (Dicken and Lloyd 1981; Short 1981; Streit 1977).

The net outcomes of all these changes in the geography of employment clearly represent a fundamental influence on many other dimensions of social geography. A full examination of the geography of employment change would merit a book in its own right; but it is nevertheless important to establish at least the broad outlines of contemporary patterns. Three components of the overall picture are briefly examined here: manufacturing employment, female activity rates and unemployment.

Regional Variations in Manufacturing Employment. As we have seen, there has been, since the 1960s, a general trend towards the relative decline in the manufacturing labour force in Western Europe. As Figure 5.1 shows, however, this trend has affected different countries at different rates, thus changing the relative share of the diminishing pool of manufacturing jobs. Unfortunately, there is little by way of evidence of such differentials at the *regional* scale across the whole of Western Europe. Some indication of the net outcome of regional changes in manufacturing employment can be gained, however, from the results of analyses of the postwar experience of the United Kingdom (Fothergill and Gudgin 1982, 1983; Keeble 1976, 1980). In simple terms, the overall decline in manufacturing employment in the United Kingdom has taken place at the expense of the 'core' regions, while peripheral regions have begun to experience an absolute gain in manufacturing employment. The early postwar period was one of massive growth, focused on the London and Birmingham conurbations. Until 1966, as the national pool of manufacturing jobs expanded, only a minority of sub-regions experienced a net loss of manufacturing jobs. Among these, London and the older industrial conurbations of Manchester, West Yorkshire and Clydeside were most prominent. After 1966 – the year in which national employment in manufacturing began to decline – decline was experienced in nearly all the major industrial centres; meanwhile, peripheral sub-regions experienced a net gain, with a 'consistent and remarkable gradation' (Keeble 1980: 947) 'from a high rate of decline in the conurbations through a lower but still considerable decline in the more urbanised counties, to only moderate decline in the less urbanised counties and actual growth in rural areas'. The available evidence suggests that a similar pattern of decentralisation has been occurring throughout Western Europe, though the overall pattern remains to be established in detail. Nevertheless, it is clear that the recent overall decline in manufacturing employment has been associated with a significant shift in the balance of employment opportunities.

Variations in Female activity Rates. Sectoral economic change and corporate reorganisation, particularly the shift in emphasis from manufacturing to service employment and the proliferation of branch plants within the manufacturing sector, have resulted in an increase in the proportion of economically active women in every country of Western Europe. The amount and the rate of increase in female activity rates, however, has varied a good deal, partly because of regional differentials in these changing patterns of demand for labour, and partly because of regional differentials in the propensity for women to enter the labour market, or to re-enter the labour market after a phase of child-rearing (which generally lasts for about ten years in northwest Europe but longer in Mediterranean Europe) (International Labour Office 1974). In Ireland, for example, female activity rates have remained relatively stable, rising from 19.7 per cent in 1970 to 20.1 per cent in 1980; while in Denmark the increase over the same period was from 36.9 per cent to 45.2 per cent. Once again, it is difficult to assemble reliable regional data for the whole of Western Europe. The interregional gradient in female-activity rates can be illustrated with reference to the nine nations of the European Community in 1980, however (Figure 5.2). Within this area, female activity rates ranged from less than 15 per cent in Friesland (Netherlands), Sicily and Sardinia to over 40 per cent in Denmark and Île-de-France. The most striking feature of this map, however, is its apparent lack of correlation with patterns of economic structure or prosperity. High activity rates are found in *some* peripheral regions and *some* metropolitan, core regions; so are low activity rates. What this reflects, of course, is the mediation of patterns of economic opportunity (stemming from economic structure and corporate organisation) by social, demographic and cultural factors. This, in turn, serves to emphasise the multidimensional impact of changing female activity rates in different regional settings. As McNabb (1980), Wainwright (1978) and others have pointed out, the economic emancipation of women and the extra incomes associated with increases in female activity rates have to be set against possible retrenchment in male employment opportunities, the continued sexual division of labour (in which the range and quality of female employment opportunities is generally inferior), and the repercussive effects on culture and family organisation. Figure 5.2 thus represents a highly important but under-researched dimension of regional social geography.

Regional Variations in Unemployment. Changes in the rate of unemployment are less ambiguous, although standardised data are equally difficult to obtain at the regional level for Western Europe as a whole. Once again, therefore, we refer to the experience of European Community members. Figure 5.3, which illustrates the changing totals of unemployment by nation between 1971 and 1982, shows the dramatic increase in unemployment which followed the 1973-4 oil crisis. From a level of less than 3 million in 1971, when one of the major concerns of an International Labour Office report was the increasing difficulty of filling unpopular jobs, total unemployment had risen to more than 10 million at the end of 1981, with unemployment in the United Kingdom alone standing at over 3 million. The regional patterns associated with these changes are shown in Figure 5.4. Quite clearly, the effects of recession, structural change and economic reorganisation have been highly differential in their impact on unemployment rates. The worst change, in *relative* terms, occurred in the industrial core regions of northeast France and western Belgium, where unemployment rates surged from between 1 and 2 per cent in 1973 to between 7 and 10 per cent in 1980. Other regions which experienced a marked relative deterioration during this period include several which have for a long time been regarded as leader regions (the West Midlands and East Anglia in England, Hamburg and Nordrhein-Westfalen in West Germany, and the Paris Basin in France) as well as some of the 'traditional' industrial heartlands (Saarland in West Germany, northern England, Northern Ireland and Wales) and peripheral rural regions (Campania and Sardinia in Italy, and Languedoc-Roussillon and Auvergne in France). On the

Figure 5.2: Female Economic-activity Rates, 1979

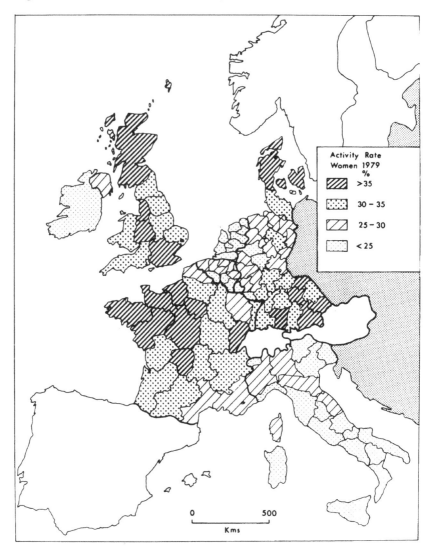

Source of data: National statistical abstracts.

Figure 5.3: Changes in Total Unemployment, 1972-81

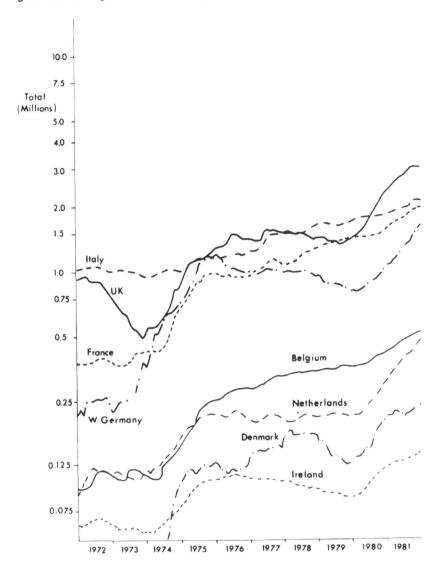

Source of data: National statistical abstracts.

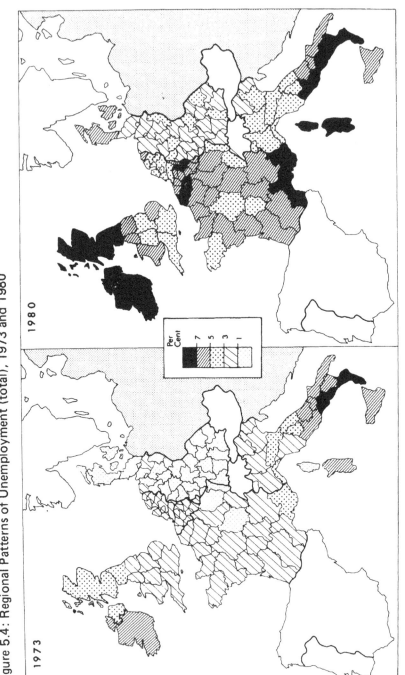

Figure 5.4: Regional Patterns of Unemployment (total), 1973 and 1980

Source of data: National statistical abstracts.

other hand, the relative deterioration in unemployment in Ireland and much of southern Italy has been somewhat less pronounced. Nevertheless, these areas remain firmly at the foot of the unemployment league table, with a steep gradient separating them from the best-off regions. In 1980 the five worst-off regions within the European Community were Calabria and Campania (Italy), Liège and Hainaut (Belgium) and Sardinia, with unemployment rates of around 10 per cent, four times. the rate experienced in the best-off regions (the Duchy of Luxembourg, and Stuttgart, Karlsruhe, Freiburg and Oberfranken in West Germany). Such patterns, of course, represent a significant component of the contemporary social geography of Western Europe. In this context, it should be borne in mind that the overall figures depicted in Figures 5.3 and 5.4 inevitably obscure regional variations in the intensity of the various components of unemployment – youth, long-term, short-term, structural, cyclical, and so on – which effectively stratify unemployment in different ways in different locations, leading to different types of social problems. This, however, is another area which has been under-researched at the regional level.

Centres and Peripheries

The cumulative effect of the differential impact of successive waves, phases and cycles of economic development has often been interpreted in terms of *centres* of capital accumulation and economic power and *peripheries* of limited – or suppressed – potential for economic development (Commission of the European Communities 1981; Despicht 1980; Folmer and Oosterhaven 1979; Gottmann 1980; Lewis and Hudson 1984). In terms of economic activity there is certainly a sharp gradient between, on the one hand, the major urban/industrial complexes of northwestern Europe and, on the other, the rest of Western Europe. Moreover, the gradient appears to be steepening. This has been clearly demonstrated by Keeble, Owens and Thompson (1982) in an analysis of variations in 'economic potential' among the regions of the European Community. Their index is a measure of the 'centre of gravity' of economic activity, based on regional GDP and taking account of interregional transport costs and tariff barriers. As such, it can also be interpreted as a measure of regional comparative advantage for economic growth. Figure 5.5 shows that the current pattern is characterised by a triangular plateau of high potential in the northeast of the Community, with outlying secondary peaks of relatively high potential around Paris, London and West Berlin and extensive areas of low potential to the southern, western and northern fringes of the Community. This pattern represents a relative deterioration in the position of the peripheral areas — a trend which is not, apparently, the result of the entry of successively poorer nations to the Community. Rather, it seems to be related to intensifying centre/periphery differences in regional economic structure. Thus, while both peripheral and central regions recorded falling proportions of employment in manufacturing and rising proportions in service industries during the 1970s, the decline in the periphery's already lower manufacturing proportion was in fact faster than that for central regions. Moreover, its growth in *producer services* (transport, telecommunications, finance, banking and insurance) was relatively slow, while its growth in *consumer services* (education, health, public administration, retailing, etc.) was relatively fast (Keeble, Owens and Thompson 1981).

Further evidence of the dominance of the northeastern 'centre' of the Community is provided by Figure 5.6, which shows the pattern of regional productivity (GDP/worker) measured in relation to the Community average. Once again there is a striking gradient, with productivity in the 'best' regions over 50 per cent above the Community average, compared with levels of 50 per cent or less in parts of southern Italy. Once again, this represents a relative deterioration of the position of many of the peripheral areas. Most dramatic has been the turnaround in the position of the industrial regions of the United Kingdom, all of which had levels of productivity in 1950 which were about 20 per cent above the overall level for the nine nations. As Figure 5.6 shows, all are now more than 20 per cent below the overall level. Meanwhile, the likes of Nordrhein-Westfalen, Rheinland-Pfalz and Baden-Württemberg (in West Germany) have moved into a dominant position from a below-average position in 1950; and the adjacent regions in Belgium and the Netherlands have consolidated their above-average position to one of pre-eminence (see Molle, van Holst and Smit (1980) for a detailed breakdown of changes in regional GDP per worker since 1950).

A third useful perspective on the dominance of a centre/periphery structure within the European space-economy is provided by Erlandsson's (1979) work on the 'contact landscape'. Erlandsson argues that the increasing internationalisation of capital, coupled with the increasing importance of the quaternary sector, means that the potential for maintaining face-to-face communication between employees of both businesses and governments has become a critical determinant of regional development. Measuring the average amount of time which can

Figure 5.5: Economic Potential within the European Community, 1977

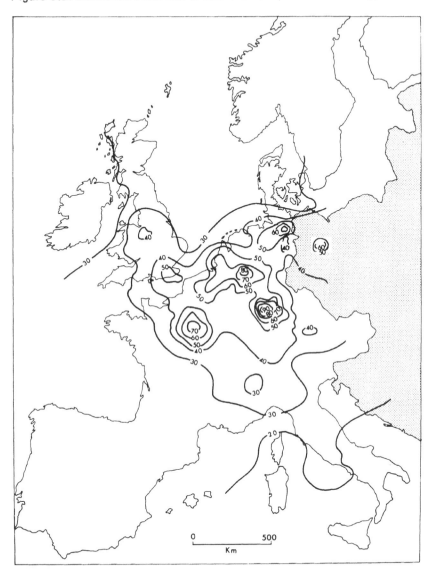

Source: Keeble, Owens and Thompson (1982).

Figure 5.6: Regional Productivity: GDP Per Employed Person, 1980

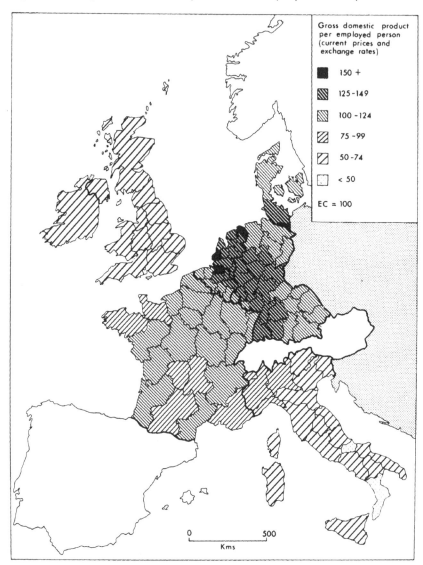

Source of data: National statistical abstracts.

be spent in other European cities within a 24-hour round trip from a given point of origin (Figure 5.7), Erlandsson is able to identify what he calls the Primary European Centre – the region delimited by lines drawn between Paris-London-Hamburg-Munich-Milan-Lyon-Paris. Outside this area, cities in Scandinavia, the United Kingdom and Ireland, and southern and western France form three relatively homogeneous regions with intermediate accessibility to other cities: they are Intermediate European Centres. The only city outside these intermediate regions with a comparable outbound 'stay time' is Madrid; the rest of Western Europe, together with all of Eastern Europe, is categorised as consisting of Secondary European Centres. This pattern is, of course, self-perpetuating. Originally the result of relative location, it is intensified by the evolution of air and rail-passenger timetables which reinforce the initial advantage of the most favourable locations.

Economics, Politics and Culture

On the basis of this evidence it is hard to dispute the notion of centre/periphery contrasts as the dominant feature of the Western European economic landscape. Yet it is also clear that different criteria yield different definitions of 'centre' and 'periphery' (see also Claval 1980). King (1982), noting the frequency of references to centre/periphery patterns in the economic geography of Western Europe, draws attention to the variety of competing definitions of the core. Since Delaisi wrote of an 'inner' and an 'outer' Europe in 1929 the core area has been interpreted as the triangular region defined by Lille-Bremen-Strasbourg (the 'Heavy Industrial Triangle'), as an axial belt stretching between Boulogne and Amsterdam in the north and Besançon and Munich in the south, as a T-shaped region whose horizontal axis runs from Rotterdam to Hannover, and whose stem extends down the Rhine to Stuttgart, and so on. Such definitions may be confusing in their variability, but their major weakness is that they overlook the interdependence which exists between centre and periphery. It is therefore more satisfactory, as King (1982) points out, to think in terms of a centre consisting of a number of *structural* components (demographic, economic, welfare systems) which are linked to the structural components of peripheral regions by a series of *flows* (capital, migrants, taxes, tourists, social fashions) which bind centre and periphery together, reinforcing their symbiotic relationship.

Seers (1979) makes the point that there are in this sense two 'centres' to the European periphery: Europe's own dominant centre and that of the United States. Both are suppliers of capital and technology,

Figure 5.7: The Accessibility of European Cities to One Another. The values are expressed relative to the most generally accessible city (Paris = 100)

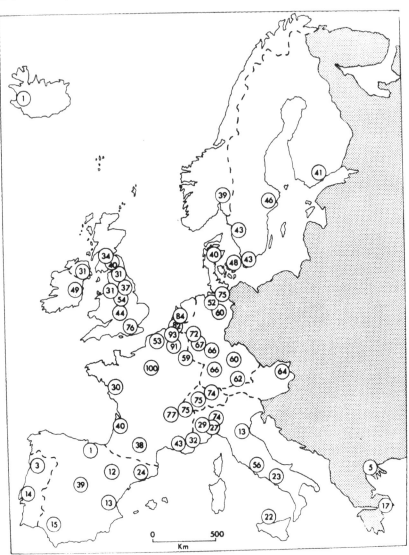

Source: Erlandsson (1979).

providers of tourists, absorbers of migration and exporters of key cultural ideas and artifacts (movies, magazines, etc.). Using these more dynamic criteria, Seers defines the European centre as an incomplete egg-shape centred on Kassel in West Germany, and with its long axis stretching 2,700 km from Barcelona in the southwest to Helsinki in the northeast (Figure 5.8). Generally, 'the closer to the centre of the egg, the greater the concentration of power. One could imagine a "yolk" which would include Denmark, West Germany, the Benelux countries, Paris, the Lyon area, Switzerland and Lombardy' (Seers 1979: 19). Moreover, most of the peripheral countries have their respective national cores orientated towards this 'yolk'. On the other hand, the symmetry of the pattern is disrupted by the United Kingdom, Italy and Finland, which Seers classifies as 'semi-peripheral'. The United Kingdom, for example, exhibits socio-economic characteristics which are not noticeably different from those in the centre, yet it is

a country increasingly penetrated by foreign capital, which has gained a big fraction of the North Sea oil concessions, and by foreign goods, which have captured a large share of important domestic markets . . . It is now a net receiver of tourists. (pp. 15-16)

And, like Italy, it is weak in government R and D activity, debilitated by chronic inflation, and dogged by periodic foreign-exchange problems.

Seers's use of cultural as well as economic criteria is a useful pointer to the political economy of centre/periphery relationships. As Gottmann (1980) has noted, the existence of a centre/periphery structure not only implies the functional organisation of space but a degree of conflict between the dominant centre(s) and subordinated peripheries. Economic dominance, in other words, tends to go hand in hand with political control and cultural standardisation. This, in turn, points to the existence of different *kinds* of peripheral regions which have resulted from the interplay of economic, political and cultural factors. In this context, the work of Stein Rokkan and his associates is particularly useful. Rokkan himself (1980) demonstrated how the emerging patchwork of nation states (see Chapter 3) created a number of peripheries as *interfaces* between major politico-economic core territories. These include parts of Seers's centre such as Flanders/Wallonia, Alsace-Lorraine, the Bernese Jura, Savoie and the Austro-Hungarian interface in the Burgenland (Figure 5.8). He also recognised *external* peripheries, characterised by their relative distance from national centres. These include Brittany, Galicia, Scotland and Wales (Figure 5.8). At the same time, differentiation along the major lines of cleavage between centre and periphery – political, cultural and economic – are recognised as important. In terms of economic resources, for example, peripheral regions can be grouped into *industrial peripheries* (e.g. northern Italy, northern England), *service peripheries* dependent on tourism (the French southern Atlantic and Mediterranean coast, the Italian Riviera, south-western England) and *deprived peripheries* (southern Italy, the Republic of Ireland, Scotland, Wales, Sardinia and Corsica) (Figure 5.8). Thus, there emerges the potential for a sophisticated typology of centres and peripheries. This is what Rokkan and his successors have been working towards (Aarebrot 1982; Rokkan and Urwin 1982), and because it seeks to combine the outcomes of political, cultural and economic processes within a territorial framework, it represents a potentially fertile approach for regional social geography.

Regional Policies for Industrial Europe

It was the depression of the 1930s which first prompted state intervention on behalf of depressed industrial regions in Western Europe. In Italy and Germany, fascist governments relied heavily on public works and infrastructural programmes in their attempts to rejuvenate the worst-hit regions; elsewhere, particularly in Sweden and the United Kingdom, governments also attempted to assist distressed industrial areas via public works and public expenditure, though they preferred to model their efforts on the example of the American New Deal experiments. It was not until the postwar period, however, that regional policies came to be of any real significance to the social and economic well-being of industrial Europe. Nicoll and Yuill (1980) have identified three major policy phases, each reflecting different circumstances and constraints:

(1) a period of tentative policy development from 1945 to the late 1950s;
(2) a period of innovative and very active regional policymaking which lasted to the early 1970s;
(3) a muted phase, from around 1973 to the present.

It must be remembered, however, that the development of regional policies in individual countries has, in detail, produced a complex variety of policy instruments whose origins, goals and consequences are not always easy to identify. There is a large literature on this subject. Comprehensive reviews are given by Hudson and Lewis (1982), Kuklinski (1981), OECD (1976a, 1976b), Pinder (1983), Vanhove and Klaasen (1980) and Yuill, Allen and Hull (1980); useful reviews of national experience include those by Berentsen (1978) on Austria, Aydalot (1978) and Estrin and Holmes (1983) on France, Chiotis (1972) on Greece, Bianchi (1979) on Italy, de Smidt (1979) on the Netherlands, Lopes (1981) on Portugal, de Terán (1981) and Richardson (1975) on Spain, Dunford, Geddes and Perrons (1981), House (1982) and Moore (1980) on the United Kingdom and Zielinski (1983) on West Germany. What follows is a brief review of the nature and impact of regional policy during each of the major policy phases.

Figure 5.8: Centre and Periphery in Western Europe.

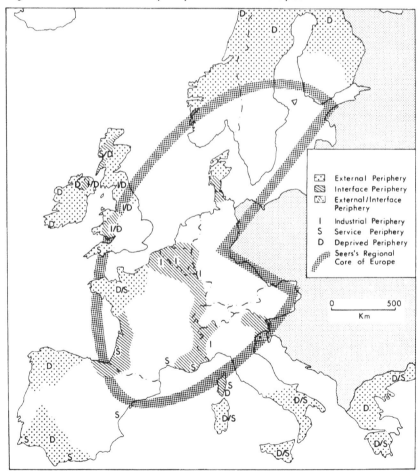

After Seers (1979) and Aarebrot (1982).

The First Policy Phase: Tentative Beginnings

During the first ten to fifteen years after the Second World War, most countries in Western Europe established tentative policy frameworks whose objectives were focused on 'taking work to the workers' in depressed, industrial regions. In general terms, the motivation for this activity was largely based on the Keynesian view that governments had both the responsibility and the ability to counteract regional recession, which was seen as an obstacle to the most efficient use of national resources. For the most part, however, such policies were not pursued very vigorously: it was, after all, a time of recovery and reconstruction throughout Europe, when the emphasis was firmly on national growth and well-being with little interest in the 'luxury' of its spatial distribution, and when many of the traditional industries − coalmining, shipbuilding, iron and steel, textiles − were experiencing a minor boom because of the backlog of demand caused by the war. In Italy, the Cassa per il Mezzogiorno did represent a major regional-policy initiative; but until 1957 its resources were overwhelmingly oriented towards the agricultural infrastructure (Borzaga and Goglio 1981). Only in the United Kingdom was a significant effort directed towards industrial regeneration, and then only for a limited period. The postwar Labour Government revived the policy measures of the prewar Special Areas legislation, which included government factory-building in designated Development Areas, loans to industrial estate companies, and discretionary grants or loans to approved commercial projects (Lonie and Begg 1979; Pitfield 1978). In addition, the system of Industrial Development Certificate (IDC) control was introduced in 1948. Under this system, all new projects above a certain minimum size were required to obtain an IDC in order to be granted planning permission. Before the system could gather any real momentum, however, an external balance-of-payments crisis forced the government to downgrade the policy and, with the incoming Conservative Government of 1950, regional policy found itself placed firmly in the background until much later in the decade (McCrone 1969).

The Second Policy Phase: Expansion and Innovation

After the late 1950s, the problems of depressed regions − both rural and industrial − became more acute as long-term structural change was compounded by a series of minor recessions. This, in turn, prompted the 'rediscovery of poverty' and the emergence of a growing concern over questions of equality and distribution, particularly in the regional context where, as Hoffman (1977) emphasises, economic differentials interacted with ethnic and cultural differentials. At the same time, migration from economically marginal and geographically peripheral regions contributed to serious problems of congestion and infrastructure provision in major agglomerations, particularly London, Paris, the *Randstad* and Milan-Turin-Genoa. 'In such a situation,' write Nicol and Yuill (1980),

> the case for regional policy was strong, and it was further strengthened by a general confidence at the time in government intervention; a conviction in many parts of Europe that the problems facing government could be solved if only appropriate planning was undertaken and the correct policies adopted. (p. 19)

Moreover, regional policies were able to gain general support and legitimacy because they were presented as serving the interests not only of peripheral regions but also of metropolitan, leader regions and, indeed, of entire national economies: rural development and industrial regeneration were to be complemented by deflationary constraints in metropolitan regions; while both were seen as contributing to 'balanced' national economic management.

The policy measures which emerged in response can be grouped into four main categories. First, there was a continuation of *infrastructural* schemes; though by the mid-1960s it was widely recognised that such schemes were 'permissive' of industrial development rather than 'propulsive'. Nevertheless, infrastructural investment of all types has been, and remains, (particularly in the more peripheral countries) a key element of regional development. National economic growth requires the transformation of the environments of economically marginal regions by modernising them and creating, through modern road systems, hospitals, houses and technology parks, new settings that are more attractive to private capital.

Secondly, there was the development of state-owned industry for regional development purposes. As we have seen, this became a major feature of the Italian regional strategy after 1957. Most other governments have acquired some interest in manufacturing industry, however, and this has inevitably led to pressure to use the state sector to the benefit of problem regions. Similarly, several countries have begun to decentralise the administrative component of government, moving civil servants to provincial cities (see also Pacione 1982b).

Thirdly, there were a variety of disincentives and controls designed

to curb development in core, metropolitan regions; or at least to
encourage entrepreneurs to consider locating in a peripheral region. The
British system of IDC control was considerably strengthened in 1965,
when similar controls were imposed on the office sector through a
system of Office Development Permits. In France locational control in
the Paris region through a system of permits (the *Agrément*) was among
the first regional policy measures in the mid-1950s. Applied initially to
the manufacturing sector, its coverage was widened to include the
office sector in 1958, and it was augmented by a type of congestion tax
– the *Redevance* – in 1960. Italy, too, developed a mixed system of
administrative control and fiscal penalty, while in the Netherlands
permits are necessary for new industrial development in the Rijnmond
area, and levies are payable by industries operating in the 'control zone'
covering the provinces of North and South Holland, Utrecht, and
Gelderland.

The fourth category of regional policies consists of incentives.
Because they are an essential complement to other categories of policy
and because they account for substantial amounts of public expenditure, they have come to be regarded as the cornerstone of regional
policy in most countries. Regional incentives aim to encourage industry
to locate or expand in designated marginal regions by lowering the costs
of particular factors of production (usually capital). Influenced by a
succession of vogue theories in regional economics and regional planning, governments developed a wide and complex variety of incentive
systems during the prosperous mid-1960s, when the necessary budgetary
allocations were relatively easy to obtain. The most widespread types of
incentives were capital grants, tax concessions, interest-related subsidies,
depreciation allowances and labour subsidies, with most countries
offering a 'package' of two or three key incentives in designated regions
(Nicol and Yuill 1980).

Common to all four categories of policy throughout this second
phase was the tendency to designate ever larger portions of national
territory subject to regional policy. In part, this can be attributed to the
relative affluence of the period, but it was also the result of political
considerations and of a growing awareness of the variety and complexity of the 'regional problem' which the policies were seeking to
address. In terms of the targeting of regional policy, however, the
dominant feature of the period was growth-point strategy, which
seemed to offer the advantages of localisation and agglomeration
economies, and the potential for the creation of self-sustaining growth
through local cumulative causation processes (Moseley 1974; Richardson

1969). In addition, growth-point strategies were attractive because they made for a more 'efficient' use of public funds for infrastructural investment. French regional policy is probably the best example in this context. Eight major growth points (*métropoles d'équilibre*) were designated in 1964 in an attempt to reduce the centre/periphery gradient which Gravier (1947) described as 'Paris et le désert Français'. Thanks to a well-funded package of policy measures, these métropoles (Lille, Metz-Nancy, Strasbourg, Lyon-St Étienne-Grenoble, Marseille, Toulouse, Bordeaux and Nantes-St Nazaire) were able to expand rapidly during the latter part of the 1960s. At the same time, however, the problems of growth-point strategy also began to emerge. These included the creation of regional centre/periphery situations and the relative costliness of funding development in the métropoles in comparison with smaller centres. Moreover, it became increasingly apparent that the métropoles were unlikely to be able to modify the dominant trend of centralisation focused on the Paris region, and that many industrial firms were in any case displaying a preference for small towns rather than large urban centres (Prud'homme 1974; Tuppen 1980). Similar problems were encountered in other countries where growth-point strategies had been pursued (Breathnach 1982; Brown and Burrows 1977) so that, by the end of this policy phase, their effectiveness came to be widely questioned.

As with all strategies and policy measures, however, it has proved difficult to establish degrees of success. For one thing, regional policies rarely incorporate specific objectives. More important, regional policies have not been the only force affecting the 'trajectory' of regions in economic space. The effects of sectoral economic change and changes in spatial organisation – particularly the increasing internationalisation of capital and the international division of labour – are very difficult to isolate, analytically, from those of regional policies. Moreover, the methods employed in policy evaluation – shift-share analysis and linear modelling – are themselves difficult to operationalise effectively (Nicol 1982). As a result, it is not surprising to find that the conclusions of evaluation exercises can vary a good deal. Estimates of the number of jobs created in the United Kingdom by regional policies between 1960 and 1973, for example, range between 11,000 and 72,000 per annum (Buck and Atkins 1976; Mackay 1976; Moore and Rhodes 1975; Schofield 1976). Ashcroft (1980), reviewing attempts to evaluate regional policies in Belgium, Denmark, Ireland, the Netherlands, West Germany and the United Kingdom, suggests that regional policies in these countries can generally be considered to have induced between 10 and 20 per cent of all new investment in their designated areas — the equivalent of about 1.2 jobs per annum per thousand population in the designated areas. Such figures demonstrate that it is possible to generate employment in marginal areas. Moreover, the consensus to emerge from evaluation studies is that it is also possible to achieve a degree of decentralisation from leader regions; and that the costs of regional policies need not outweigh the benefits for a national economy. What is less clear, however, is the extent to which regional policies have been able to narrow the gap in productivity, incomes and opportunities between leader regions and the rest. The feeling among many analysts, however, is that the regional policies of the 1960s and early 1970s were disappointing in their results, at best enabling distressed regions to 'run in order to stand still' (Hall 1982).

The Third Policy Phase: Policy in an Age of Recession

The policy environment changed markedly as a result of the oil crisis of 1973-4. No longer was 'overheating' and congestion in leader regions perceived as a problem; no longer was unemployment and under-employment limited to traditional problem regions; and no longer was economic growth taken for granted (Cox 1982; Dahrendorf 1982). The view that the essential task was one of regional management against a background of economic growth quickly evaporated. Likewise, the accompanying theoretical and methodological tool-kit no longer appeared relevant. Meanwhile, the imprint of corporate reorganisation, the internationalisation of capital, and the sectoral shifts associated with late capitalism all became more explicit, creating new macro-economic management problems for governments, and significantly changing the nature of regional problems. Not only did existing regional problems deepen and intensify, but new ones also emerged, some of them, as we have seen, in regions that had previously been in the van of national economic growth. In this context, the regional policies of the 1960s, which were aimed at subsidising companies' fixed-capital investment costs, came to be seen as having reinforced the process of external control through the creation of branch-plant economies (Damette 1980; Hudson and Lewis 1982).

In this more hostile environment, regional policy was soon down-graded. In the face of the daunting combination of economic recession and high inflation, regional policy budgets have been trimmed, and the management of regional problems has been seen increasingly as dependent upon *national* economic policy (Allen 1981; Moore and Rhodes 1982). The most dramatic cutbacks to regional policy have been in the

United Kingdom, where the election of a Conservative government in 1979 was followed by an immediate decision to cut regional policy expenditure by 40 per cent, reducing the magnitude of assistance, reducing the extent of the designated areas, and relaxing the constraints on development in non-designated areas. Denmark and West Germany soon followed with similar, though less severe, cutbacks (Allen 1981). Elsewhere, retrenchment and consolidation, rather than cutbacks, have dominated regional policy. Many fiscal incentives have lost their attractiveness because of the generally lower levels of taxable profits throughout Western Europe; and most countries have relaxed disincentives and controls on both industrial and office location as their rationale has been removed by the intensifying social and economic problems of inner-city areas within leader regions.

Not surprisingly, the period has seen few significant innovations in regional policy, notwithstanding the proliferation of regional government and regional development agencies. Nicol and Yuill (1980) point to the emergence of more selective regional aid and the introduction of small, 'gap-filling' micro-policies, while Allen (1981) suggests that the search for new targets for regional policies has resulted in a shift away from the idea of influencing the locational decisions of larger manufacturing enterprises towards the idea of encouraging the indigenous regional development of small firms in general and the service sector in particular. The Irish Government, for example, has introduced an employment grant for jobs created in certain services (e.g. data processing, computer software development, commercial laboratories, training services), while the West German Government has introduced a special investment grant relating to high-grade managerial and R and D jobs. Some governments, with an eye to the predicted upswing of the fifth Kondratieff cycle, have begun to establish enterprise zones associated with technology parks. Such initiatives are seen by Maillat (1982) and Rothwell (1982) as critical to future regional development, but as Hall (1981a) observes, it is optimistic to think that it will be possible to create the right climate of innovation in the depressed, peripheral regions of Europe.

The one major regional policy initiative to emerge in the post-1973 period was one which, ironically, was rooted in the economic and political climate of the 1960s: the European Regional Development Fund (ERDF) of the European Community. Although the Community had effectively operated limited 'regional' policies through the European Coal and Steel Community (ECSC) and the European Investment Bank (EIB) for some time, there was, until the mid-1970s, no comprehensive,

Figure 5.9: European Community Regional Development Fund Grants
1975-8

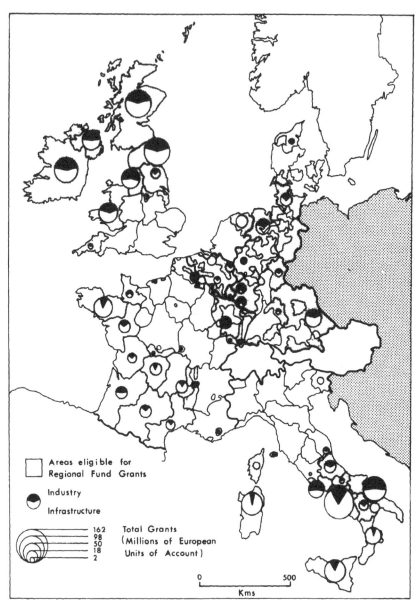

Source: Commission of the European Communities (1980b).

co-ordinated framework within which to operate (Klein 1981; Molle and Paelinck 1979; Pinder 1983). The ECSC, as we saw in Chapter 3, was limited to the 'readaption' of workers and the 'conversion' of local economies in depressed coalmining and steel-producing regions. The EIB is a Community banking system designed to reduce intra-Community disparities in economic development by disbursing loans to selected projects in priority regions (Pinder 1978); but although it has been particularly influential in sponsoring projects in marginal trans-frontier regions, it is not equipped to disburse grants or other incentives, and its performance in tackling regional development problems has been widely criticised within the Community. Moreover, by the early 1970s it was increasingly argued that the CAP, the major budgetary item within the Community, was operating to the benefit of rich nations and regions at the expense of poor regions, particularly depressed industrial regions. The United Kingdom, on entry to the Community in 1972, pressed for the establishment of a regional fund which would help to redress the balance. Eventually, the ERDF became operative in 1975 when it was allocated a relatively modest budget (1,300 million European Units of Account, compared with the 2,250 million recommended by the Thomson Report in 1973). Following the principle of 'additionality', projects must qualify for national funding in order to receive Community support: the ERDF thus claims to complement rather than replace national strategies, and preference is given to projects in regions designated as priority areas by national regional policies. At the same time, member governments are required to restrict their assistance to projects in order to ensure that financial support does not exceed a predetermined ceiling (which ranges from 20 per cent of total investment costs in the most prosperous regions to 75 per cent in the Mezzogiorno, the Republic of Ireland, Northern Ireland and West Berlin). Such controls are part of the Community's attempts at policy harmonisation, aimed in particular at eliminating 'competitive over-bidding' between nations for private capital (Swann 1983). The bulk of ERDF funds (>85 per cent) are allocated to member nations on a quota basis. In recent years about a third of this has been allocated to Italy, with the United Kingdom receiving a further 25 per cent, and France and Greece each receiving around 13 per cent (Figure 5.9). France, West Germany, the Netherlands, Belgium, Denmark and Luxembourg are all net contributors to the fund. Because of the preference given to national priority areas, this effectively channels a large proportion of ERDF funds to the Mezzogiorno, Sardinia, Scotland, northern England, Wales, Northern Ireland and Brittany (Martins and

Mawson 1980). Although encompassing most of the Community's most acute problem regions, this distribution is generally regarded as poorly targeted. The Commission of the Community has in fact wanted to target ERDF funds more explicitly from the outset, using a standardised criterion based on a combination of (i) regional GDP *and* either (ii) regional dependence on agricultural employment *or* (iii) a persistently high rate of unemployment *or* (iv) a high rate of outmigration (Thomson Report 1973; Commission of the European Communities 1977). The quota system has proven to be more acceptable politically, however. Meanwhile, other aspects of the ERDF have attracted criticism. These include the system of ceilings on national support for eligible schemes (which is felt to be unjustifiably crude in its structure), and the absence within the ERDF framework of any Community-wide system of controls and disincentives which could complement the incentives offered (Martins and Mawson 1982). Above all, however, the ERDF has disappointed because of the trifling level of its budget in comparison with the CAP (Armstrong 1978; Martins and Mawson 1981; Yuill *et al*. 1980). In conclusion, however, it seems appropriate to echo the reasoning of Holland (1976a) and others who point out that the new forms of corporate organisation associated with late capitalism are increasingly less likely to be influenced by the kind of localised incentives offered by the European Community, whatever the size of the budget. In short, the new international division of labour means that the peripheral regions of Western Europe are in direct competition with much of the Third World, where the advantages to transnational corporations may easily outweigh those represented by regional incentives in developed nations (Coates, Johnston and Knox 1977).

6 URBAN EUROPE

Although Europe did not invent urban civilisation, it was the first continent to establish a true system of cities and the first to experience urbanisation *en masse*. European urbanism has a long and distinguished history, and if there is a 'European' culture then it is first and foremost an urban culture. Indeed, the majority of people live and work in towns or cities, and much of Western Europe has reached the point of 'saturation urbanisation', with metropolitan areas coalescing to create extensive megalopolitan regions. Assessing the true extent of urbanisation is not easy, since national definitions of what actually constitutes an urban settlement vary from a threshold of 200 (in Norway and Sweden) to 20,000 (in Greece and Spain). But, according to individual national definitions of 'urban' locations, only in Norway and Portugal do less than half the population live in towns and cities; in Belgium, Denmark, France, Iceland, the Netherlands, Sweden, West Germany and the United Kingdom the proportion ranges between three-quarters and four-fifths.

The framework of the present urban system was forged by the end of the Middle Ages. The embryo urban system introduced by the Greeks and extended by the Romans almost collapsed during the Dark Ages, though many of the survivors – including Athens, Köln, London, Paris and Naples – did go on to become major cities. The mercantilist era of the Middle Ages, however, required an extensive hierarchy of towns and cities and, for the most part, the locational logic of these trading settlements has ensured their continuity. Successive waves, phases and cycles of economic development have, of course, modified the relative size of some cities and transformed the function of most, while some new cities have been created and a few have declined. The first important differentials in urban development occurred during the seventeenth century, when the trading advantages enjoyed by Atlantic and North Sea ports gave them an impetus which could only be matched by other settlements if they were fortunate enough to have become princely residences, naval stations or centres of absolutist administration (de Vries 1976). The 'interior' towns and cities of continental Europe, meanwhile, expanded at a more modest rate, with inter-urban differentiation being determined mainly by relative accessibility to major trade routes (Diederiks 1981). When the industrial revolution came to Europe in the nineteenth century, the most dramatic changes were not so much in terms of inter-urban differentiation as in the unprecedented proportion of the population which was drawn to the cities. There were of course some components of the urban system which were not suited to the economic logic of industrial capitalism, and which were therefore bypassed by the urban transition of the nineteenth century. Many of these remain within the present urban system as small, quiet market towns – picture-postcard places like Bernkastel, Caserta and Dorchester, which give character and flavour to the European urban scene. At the same time, the industrial revolution saw the genesis of some towns: the coalfield towns of South Wales, Durham and the Ruhr, for example, and the textile towns of Lancashire, West Yorkshire and Belgium. But, in general, because industrialisation came after the introduction of railway networks (the major exception here was the United Kingdom), industry tended to locate in the existing cities which had, logically, become the major nodes of the new transport network (see, for example, Pumain 1982).

Since 1900 this basic framework has been relatively stable, with the overall degree of urbanisation tending (until very recently) to increase, while individual cities have moved up or down the urban hierarchy in response to local economic fortunes (Boudeville 1978; van der Haegen 1983). The main reason for the intensification of the urban system was, of course, the very high rate of migration from the countryside to towns and cities, first in response to the growing demand for industrial labour and later as a consequence of the rationalisation of landholdings and agricultural methods. Later still, during the 1950s and 1960s, the growth of European cities was fostered by the relatively buoyant rates of natural increase in the population (Hall and Hay 1980). With the intensification of the urban system came the growth of large cities. Clout (1975) has noted that the number of cities with more than 100,000 inhabitants grew from 200 in 1950 to 300 in 1970, accounting for 27 and 35 per cent respectively of the total population of Western Europe. Similarly, the number of 'millionaire' cities grew from 13 in 1930 to 26 in 1980. At the top of the urban hierarchy some of the largest urban regions began to coalesce into polycentred megalopolitan regions. The exemplars are the Rhine-Ruhr agglomeration and Randstad Holland. The former encompasses an area roughly 110 km in diameter, containing 22 cities with a total population of approximately 11 million (Hall 1977). Randstad Holland is a horseshoe-shaped region with a diameter of roughly 100 km, stretching from Eindhoven through Dordrecht, Rotterdam, The Hague, Haarlem, Amsterdam, Utrecht and

Nijmegen. These two megalopolitan regions, linked by the E36 motor-way, are only 100 km apart, so that it is very likely that they will eventually join up to form a single European metropolitan core. Mean-while, across the English Channel, the six conurbations of England (Greater London, Merseyside, southeast Lancashire, Tyneside, West Midlands and West Yorkshire), with a total population of 15 million, jointly represent the kind of extensive urbanisation which is only paral-leled outside Britain by the Boston-Washington axis in the USA.

Within the rest of the urban system, patterns of growth and decline have broadly reflected the regional economic changes described in Chapter 5. In addition, the fortunes of some cities (Berlin, Vienna) have been markedly affected by changes in political geography. More recently, the economic integration of Western Europe has affected the develop-ment of national urban systems. Thus, because of the increasing interna-tionalisation of capital and the removal of tariff barriers, urban popula-tion has rapidly increased in the regions situated at the border between Switzerland and Italy, in the Dutch regions at the border with West Germany, and in the French, Swiss and West German regions of the southern Rhine (Drewett and Rossi 1981). And, until the early 1970s, the movement of migrant labour contributed significantly to urbanisa-tion in the Benelux countries, France, Switzerland, West Germany and the United Kingdom (see pp. 37-9). So, although there has been continuity in the *framework* of European towns and cities, one of the main features of European urbanisation is its *diversity* in terms of growth and development.

The Urban System: Recent Patterns of Growth and Decline

A number of recent studies have detailed the complex changes which have occurred within the European urban system during the postwar period (Hall and Hay 1980; Klaasen, Molle and Paelinck 1981; Pumain and Saint-Julien 1979; van den Berg, Drewett, Klassen, Rossi and Vijverberg 1982). All of them point to the occurrence of a slowing down of growth rates in larger metropolitan areas as a decentralisation process from city to suburb has set in. The major cities of northern Europe were all report-ing heavy population losses by the mid-1970s, and there was even evid-ence of a radical slowing-down in the growth-rates of larger cities in southern Europe (Vining and Kontuly 1978). Hall and Hay (1980) con-clude from an analysis of fifteen European countries that a 'clean break' in urbanisation trends occurred during the late 1960s and early 1970s:

The urban cores, which had as much as two-thirds of net population growth in the 1950s, had less than one-half in the 1960s – and a negligible share in the early 1970s. The [suburban] rings, conversely, took only one-third of the growth in the 1950s, over one-half in the 1960s – and the whole of the net share in the early 1970s. (p. 87)

But, as Hall and Hay note, on balance, population was still leaving rural areas for towns and cities. One reason for this is the marked increase in the growth-rates of smaller urban areas. In nine of the ten Western European countries studied by van den Berg *et al.* (1982), for example, between 70 and 100 per cent of population growth between 1960 and 1970 took place in towns with less than 50,000 inhabitants. Intermediate-sized cities also grew relatively rapidly. In Belgium, Denmark and Great Britain, cities of 50,000-100,000 grew fastest, while in Italy and Sweden the most rapid growth took place in cities of 100,000-250,000. Only in France did cities of more than 250,000 outpace smaller settlements.

The net effect of metropolitan decentralisation and the growth of small and medium-sized cities, coupled with the 'turnaround' in population trends in some rural areas (p. 40), has been summarised as *counterurbanisation* (Fielding 1982; Koch 1980; Vining 1982). It should be stressed, however, that although there is widespread (but by no means universal) evidence for a reversal of the positive *overall* relationship between settlement size and net migration which obtained until the early 1960s, the arrival of 'counterurbanisation' does not mean that the European urban system is somehow likely to be 'drained' in favour of rural regions. Rather, it portends a shift in relative importance *within* the urban system towards smaller cities and away from the saturated urbanisation of the major metropolitan regions. Moreover, systematic statistical analysis of urban growth patterns reveals important regional variations, most of them attributable to the realignment of the urban system rather than to the effects of the rural turnaround on urban populations.

In terms of comparative regional growth, Hall and Hay have shown that, up to 1975, 'the tendency was still for population to move from rural to urban areas, not the reverse. The industrial heartland was still strong, and the remoter rural areas were still losing' (1980: 228). The slowest-growing (and declining) towns and cities have mostly been those located in national peripheries: eastern Austria, northern Jutland, northern Norway and Sweden, western France, western Ireland, north-west Spain, the Mezzogiorno, Scotland, Wales, northern England and the upland periphery of West Germany, especially the regions adjacent to

Figure 6.1: The Megalopolitan Growth Zones of Western Europe.

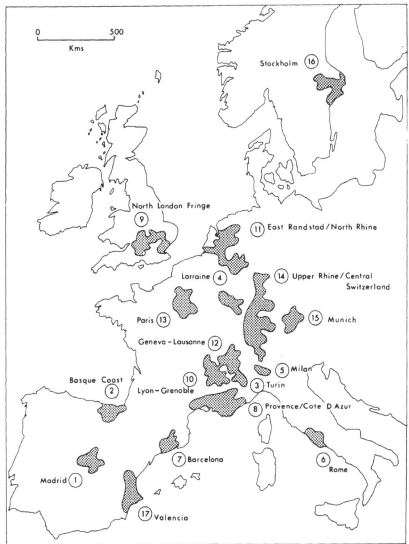

Source: Hall and Hay (1980).

Table 6.1: European 'Megalopolitan' Growth Zones 1950-70[a]

	Population		% change 1950-70	National rates of change 1950-70
	1950	1970		
1. Madrid	1,984,033	3,950,686	99.13	21.58
2. Basque Coast (Spain)	1,133,238	1,992,833	75.85	21.58
3. Turin	1,228,320	2,037,738	65.90	13.94
4. Lorraine	708,886	1,160,381	63.69	22.80
5. Milan	2,896,628	4,558,966	57.39	13.94
6. Rome	2,532,233	3,970,345	56.79	13.94
7. Barcelona	2,510,382	3,827,988	52.49	21.58
8. Provence-Côte d'Azur	2,788,690	4,242,828	52.14	22.80 (France) 13.94 (Italy)
9. North London Fringe	3,054,552	4,578,760	49.90	10.58
10. Lyon-Grenoble	1,767,740	2,548,729	44.18	22.80
11. East Randstad — North Rhine	8,690,394	12,321,128	41.66	
(German component)	3,792,987	5,383,917	41.94	19.38
(Dutch component)	4,532,534	6,383,590	40.84	30.60
(Belgian component)	364,873	544,621	49.26	14.26
12. Geneva — Lausanne — Annecy	1,031,072	1,499,659	40.59	32.35 (Switzerland) 22.80 (France)
13. Paris	7,230,690	10,068,911	39.25	22.80
14. Upper Rhine (East Bank) — Central Switzerland	9,975,364	13,855,652	38.90	
(German component)	7,684,137	10,729,651	39.63	19.38
(Swiss component)	2,291,227	3,126,001	36.43	32.35
15. Munich	1,814,585	2,508,972	38.27	19.38
16. Stockholm	1,354,434	1,828,893	35.03	14.58
17. Valencia	1,952,185	2,582,482	32.29	21.58
Total	52,653,426	77,534,951	(Av.) 47.14	(Av.) 16.82 (Europe)

Note: a. Defined as individual metropolitan regions of aggregates of contiguous metropolitan or non-metropolitan regions having growth-rates of more than 30 per cent (1950-70), and which reached one million population by 1970.
Source: Hall and Hay (1980).

Czechoslovakia and East Germany. Of the larger and more central urban regions, only London and Rhine-Ruhr have shown serious evidence of an overall slackening in growth. On the other hand, many of the fastest-growing components of urban Europe (Figure 6.1 and Table 6.1) are capital-city regions or regions which border the principal national metropolitan area (van den Berg *et al.* 1982). Also prominent among the growth zones are (i) axial regions along the Rhine, the Rhône-Saône valleys and the Côte d'Azur, where expansion has been concentrated in medium-sized cities (Hall and Hay 1980), and (ii) the major metropolitan areas of southern Europe, where rural outmigration continues to fuel urbanisation along traditional lines.

Centralisation and Decentralisation

The differential momentum of urbanisation is also reflected in regional variations in the degree of urban centralisation and decentralisation. In broad terms, capital cities and cities founded during the industrial revolution have come to experience the highest levels of decentralisation, while the larger cities of 'peripheral' Europe are still experiencing centralisation. Within Italy, for example, industrial cities like Milan, Genoa and Turin have become heavily decentralised, with some growth spilling over into smaller towns nearby; cities in central and southern Italy, meanwhile, have continued to experience significant levels of inmigration, including some from their immediate hinterland (Fielding 1982; Hall and Hay 1980). Such differentials have prompted van den Berg *et al.* (1982) to suggest that different parts of Western Europe have reached different *stages* of urban development. The first, which they call *urbanisation*, is characterised by growth and centralisation. This stage was completed, in most of the countries of Western Europe, by 1950; in a few (Denmark, Italy, Sweden) it continued to dominate until 1960. The second stage, *suburbanisation*, is characterised by the decentralisation, either relative or absolute, of population from the cores of urban regions to their immediate hinterlands, partly through higher rates of natural increase in the latter, but mainly through residential mobility. Inner-city areas thus decline in population while suburbs expand; overall, urban growth is slowed. In this context, it is worth noting that the influx of large numbers of migrant workers to the inner-city areas of Western Europe up to 1973 only softened the pace of decentralisation: otherwise, suburbanisation would have been much more pronounced (Fielding 1982; Vining and Kontuly 1978). According to van den Berg *et al.* (1982), suburbanisation has dominated most of Western Europe for most of the postwar period (Table 6.2). In some countries, however, the dominant stage is now *desuburbanisation*, where urban regions exhibit a net loss of population, with absolute decentralisation to outer suburban areas being followed by relative decentralisation as the thrust of urban growth is switched to small satellite towns and cities within 50 to 100 km. Examples include the urban regions of Amsterdam, Antwerp, Basel, Brussels, Copenhagen, Glasgow, Liège, Liverpool, London, Manchester, Rotterdam, Vienna and Zürich. In overall terms, Drewett and Rossi suggest that six of every ten households now moving into small towns (<50,000) do so within the functional urban region of their origin.

Table 6.2: Changes in Countries' Classifications by Dominant Stage of Urban Development, 1950-75

Dominant stage	1950-60	1960-70	1970-5
Urbanisation	Sweden Denmark Italy		
Urbanisation/ suburbanisation	Austria Netherlands	Sweden Italy Austria Denmark	
Suburbanisation	Switzerland Gt Britain Belgium	France Switzerland W. Germany Netherlands Gt Britain	Austria France Italy Denmark Sweden
Suburbanisation/ disurbanisation		Belgium	Gt Britain Netherlands Switzerland
Disurbanisation			Belgium

Source: Drewett and Rossi (1981).

These changes in the pattern of urban development are the product of several interrelated factors. Although there is still some debate over their relative importance, it seems clear that the fundamental influence on both 'suburbanisation' and 'desuburbanisation' (and, therefore, on 'counterurbanisation' too) has been the 'new geography' of production that has emerged in the last 20 years. As we have seen (Chapters 4 and 5), this has created a new regional division of labour as the largest cities and older industrial regions have experienced rapid deindustrialisation as a result of agglomeration diseconomies and labour problems; while rural population levels have begun to stabilise following the long process of restructuring in agriculture, smaller settlements in intermediate and peripheral regions have benefited from corporate reorganisation and the creation of branch plants, and metropolitan suburbs have benefited from the decentralisation of offices, 'new technology' industries and R and D establishments. As Fielding (1982) points out, these developments have been reinforced by the increased geographical mobility of the functionaries of large companies and central government departments. At the same time, it has been suggested that the economic recession of the 1970s intensified the process by discouraging inmigration to metropolitan inner-city areas (Vining and Kontuly 1978). Regional economic policies and regional planning strategies have also encouraged both suburbanisation and desuburbanisation in many countries. Indeed, until the comparatively recent emphasis on inner-city regeneration in the cities of northwestern Europe, practically all of the spatial policies of governments were geared towards the promotion (and channelling) of urban and regional decentralisation: urban renewal, office relocation, green belts, new towns, regional incentives and controls, and so on (EFTA 1973; van den Berg *et al*. 1982).

Another factor which has apparently reinforced the trends towards suburbanisation and desuburbanisation concerns individual environmental and lifestyle preferences. Since the urban disorders of the 1960s a good deal of attention has been paid to the disadvantages of big-city life – crime, pollution, congestion, social and ethnic conflict, and so on. Meanwhile, rural and small-town environments have become more attractive as they have been 'penetrated' by infrastructural improvements, welfare systems and employment opportunities. Ex-metropolitan households are thus able to indulge arcadian lifestyles without sacrificing the benefits of 'modern' society (hence the rural turnaround (Chapter 2) and the influx of newcomers to rural communities (Chapter 4)), or to retain an urban lifestyle within smaller towns and cities which, as Lichtenberger (1976) points out, not only offer relatively attractive environment but also a reasonable choice of employment, an adequate infrastructure and, in many cases, good facilities for shopping, education and recreation. She cites a survey by Puls (1973) suggesting that 50 per cent of the urban population in West Germany would prefer to live in medium-sized cities.

Urban Size and the Quality of Life

This raises an issue of central relevance to the social geography of Western Europe: the relationship between urban size and the quality of life. Because of differences in national census definitions and questions it is not always possible to establish the relationship very precisely with 'objective' criteria. One interesting exercise along these lines has been undertaken in Norway, where Aase and Dale (1978) ranked 9 different settlement categories (from fishing villages to major cities) according to their performance on 19 different social indicators. They found what they describe as a 'compensatory' pattern of social well-being, with larger settlements doing best in terms of incomes, working conditions and most types of services (both private and public), and smaller settlements doing best in terms of the quality of both the physical and social environment. Medium-sized settlements, meanwhile, obtained intermediate rankings on most of the 19 indicators. Similar conclusions were drawn from a survey of member nations by EFTA (1973), which suggested that well-being is probably higher in medium-sized cities on the grounds that many small settlements seem to experience serious problems stemming from low wage levels and a restricted range of job opportunities, while very large cities (>500,000) tend to experience acute social polarisation, to experience a much higher incidence of housing problems, and to be fraught with problems of access to urban amenities of all kinds – not because of limitations in the supply of opportunities, but because of the sheer size of large cities combined with the relatively high proportion of carless households. In this context it is also interesting to note that the 'Good City' identified by Donnison and Soto (1980) in their analysis of British cities was typically a medium-sized city such as Peterborough or Swindon. These cities, although they did not exhibit the trappings of elegant living or conspicuous consumption, provided a high standard of public services, and an environment in which the most vulnerable groups tended to be less disadvantaged than in other types of city.

On the evidence of people's *perceptions*, however, there is fairly clear evidence of an inverse relationship between urban size and quality of life. The fifth Eurobarometer survey, taken in 1976, provides a wide spectrum of information about poverty and life satisfaction, and most of this can be analysed by settlement size. In terms of people's overall satisfaction with life, for example, between 27 and 30 per cent of the inhabitants of settlements of less than 20,000, 20,000-99,999 and 100,000-499,999 declared themselves to be 'very satisfied' with life, compared to only 16.1 per cent of those living in cities of 500,000 or more. Similarly, residents of smaller settlements were significantly more optimistic than those living in the larger cities. Table 6.3 elaborates this overall trend by setting out the relative level of satisfaction associated with some of the major domains of life according to settlement size.

Respondents were asked to indicate their degree of satisfaction with each domain on an 11-point scale ranging from 'completely dissatisfied' to 'completely satisfied'. It should be emphasised that, in general, levels of satisfaction with most life domains were high. In every case, however, there was a significant inverse relationship between urban size and people's satisfaction. This is clearly illustrated by the proportion of respondents in each settlement category who declared themselves to be 'completely satisfied' with the various domains. As Table 6.3 shows, the steepest gradients in relative levels of perceived well-being are associated with the level of available welfare benefits, job satisfaction, transport, and people's state of health. Interestingly, the shallowest gradients (though still statistically significant) are associated with society in general and with people's idea of the level of living they feel 'entitled' to — both 'higher-order' domains which only become important issues once the more basic domains of life have been adequately satisfied.

Table 6.3: Urban Size and Perceived Well-being: Results from the Eurobarometer Survey, 1976

| | Per cent 'completely satisfied' | | | |
	<20,000	20,000-99,999	100,000-499,999	500,000+
Domain				
Housing	24.6	22.1	25.6	16.3
Neighbourhood	31.8	26.9	30.1	19.6
Household income	11.4	11.1	12.5	4.7
Consumer goods	15.2	14.3	14.8	6.3
Employment	22.0	18.6	20.0	10.2
Leisure activities	20.4	16.6	18.6	10.2
Welfare benefits				
available	17.4	14.9	12.2	3.8
Transport	24.7	21.2	19.7	14.2
State of health	31.3	24.0	28.1	20.9
Leisure time	20.2	16.4	21.9	10.1
Society in general	9.1	6.0	6.3	4.3
Personal				
entitlement	22.2	21.2	25.3	18.7

Source of data: Eurobarometer 5 (1976).

European Urbanism

The trends towards suburbanisation and desuburbanisation which have characterised the European urban system over the postwar period have gone a long way towards creating a uniformity of urban environments. Residential sprawl has affected nearly every city in Western Europe and, thanks to the economics of housebuilding and the homogenisation of consumer tastes, the very appearance of suburban houses – inside and out – has come to vary less and less. Similarly, the almost universal proliferation of renewal schemes, high-rise developments, pedestrianisation schemes, and suburban shopping malls, combined with the equally widespread *effects* of socio-economic, demographic and ethnic segregation, has made for a convergence of urban environments. European cities indeed are inching closer to the idealised 'Western' city which is more typically the product of North American and Australasian urbanisation.

Yet, such is the history of European urbanisation, a good deal of diversity persists in the townscapes, morphology and spatial organisation of cities. European cities have not only been subject to the homogenising influences of post-industrial urbanisation but also to the more selective influences of previous stages: the urbanisation of the industrial revolution, Renaissance and early modern urbanisation, medieval urbanisation and, in some cases, Roman urbanisation. Moreover, since these earlier stages of urbanisation did not occur simultaneously throughout Western Europe, the extent of the imprint of their economic organisation and political ideology tended to vary significantly from one region to another (Claval 1984).

The Spatial Foundations of European Urbanism

The richness of the legacy of European urban history is catalogued in detail elsewhere (Benevolo 1980; Burke 1976; Carter 1983; Dickinson 1951; Gutkind 1965, 1967, 1969, 1970, 1971; Mumford 1966); in this section the objective is briefly to review the major phases of urban development which contributed to the characteristics and distinctiveness of European cities up to 1950.

The earliest of these was the Roman period, the chief contribution of which was the grid layout which persists in the centre of a large number of European towns and cities. The Roman town plan was deliberately executed in response to a combination of military and religious criteria. Hygenus, a Roman architect, considered the ideal garrison town should be 725 m X 475 m, since any greater size might endanger defence by indistinct signals along the walls (Houston 1963). Inside the walls, streets were generally arranged in gridiron fashion, with their axes running north-south and east-west in accordance with religious precepts. In the North Italian Plain and the south of France the Roman legacy is very marked; one of the best examples is Turin, whose plan remained virtually unaltered within the Roman walls (which measured almost exactly 725 m X 475 m) until the urban transition of the nineteenth century and in whose centre the gridiron pattern is still clearly preserved. Other, less dramatic, examples can be found throughout the extent of the old Roman Empire, despite the morphology of many important Roman towns having been overwritten by subsequent phases of urbanisation – as in Seville, Córdoba, Toledo and Valencia, for example, where Muslim colonisation replaced grid patterns with a haphazard and tortuous street plan (Dickinson 1951).

It was in the medieval period, however, that the spatial foundations of European urbanism were laid, not just in terms of the establishment of urban hierarchies based on sophisticated trading and administrative functions but also in terms of urban morphology, urban institutions and urban culture (Pirenne 1925). Some cities emerged as ecclesiastical and/or university centres – St Andrews in Scotland, Canterbury, Cambridge and Coventry in England, Rheims and Chartres in France, Liège in Belgium, Bremen and Köln in Germany and Trondheim and Lund in Scandinavia, for example. In most cases, however, the mainspring was the mercantile economy, and it was in the merchant towns that the characteristic high-density, cellular and organic medieval morphology developed. Many modern European cities still contain fragments or lineaments of such development, while some of the towns bypassed by later phases of urban development – Mende in France, for example – are still dominated by a medieval structure and appearance. Among the best-preserved medieval urban environments, however, are those which were conditioned by defensive functions. Some, like the hilltop towns of Central Italy (e.g. Urbino, Foligno, Montecompatri) were the result of natural growth, and their morphology tended to be governed by topography. Others, like Aigues-Mortes and Montpazier in France and Vitoria and Bilbao in Spain, were planned settlements, in which gridiron plans or an elliptical arrangement of streets were common motifs.

As the political economy of Western Europe changed, so did its cities. Thus 'town plans changed from medieval diversity to baroque uniformity; from medieval localism to baroque centralism and from the absolutism of God and the Catholic Church to the absolutism of the temporal sovereign and the National State' (Mumford 1938: 77). At first, Renaissance ideology was expressed in new military fortresses laid out according to the abstract conceptions of Italian, French and Dutch theorists – Palma Nova, in Italy, is probably the best-known example. As the Renaissance developed into the baroque period, military planning gave way to the more aesthetic creation of capital cities and to the extravagant embellishment of existing towns as an expression of the despotic power of their rulers (Curl 1970). The multiplication of small towns tended to cease in most parts of Europe, and in their place national and provincial capitals grew parasitically on smaller settlements. Copenhagen, Madrid, Munich and Vienna were very much the product of this great phase of baroque town planning during the seventeenth and eighteenth centuries, and other great cities such as Bordeaux, Edinburgh, Dublin, Florence, London, Lyon, Naples, Paris and Rome felt its modifications on their existing fabric. Meanwhile, many smaller towns were established – like Karlsruhe – or modified in response to the despotism of minor princes and dukes.

Gradually, changes in world trade patterns, the rise of nation states, revolutions in military technology and the expansion of European overseas colonisation all began to redefine the internal use of urban space, thus weakening the influence of centralised planning. With the onset of the industrial revolution, however, urban structure was turned inside out as land use was redefined by the economics of industrial location, the dynamism of central business districts and the geometry of urban transport systems (Carter 1983; Knox 1982b). The spirit and ideology of this phase of urbanisation, then, are reflected by the legacy of warehouses and factories, by the symbolisation of wealth embodied in the residences and commercial properties of the new urban elite, and by the uniformity of the suburbs built by speculators for the emergent middle classes. But, while these features are widespread in their occurrence, the differential impact of successive waves and phases of industrialisation (Chapter 5) is reflected in distinctive regional variations. Thus, for example:

> Late industrialisation, the advent of railways prior to industrialisation and the coming of electric power so quickly after the 'takeoff' to industrialisation has made German cities very distinct from their British counterparts. Industries that migrated to the coalfield cities and towns in Britain remained *in situ* in German towns. The railways penetrated closer to the medieval core of German cities . . . German cities developed tram systems very early in their period of industrial growth, thus reducing the need to concentrate the workforce close to the point of employment. So the industrial colonies . . . were more 'open' than their British or Belgian counterparts. (Burtenshaw, Bateman and Ashworth 1981: 5-6)

The character of some cities was also affected by the changing geo-politics of Western Europe. After the First World War, for example, Vienna's development was arrested as it was removed from the top of an urban hierarchy with a population of over 50 million, to be left as the capital of a nation state only one-tenth the size. Soon afterwards, Dublin found itself promoted from the role of provincial capital to that of a national capital while, in Northern Ireland, Londonderry's trading hinterland was suddenly truncated.

Some Inherited Characteristics of European Cities

Running through the diversity of West European urban development are a number of distinctive attributes, rooted in past experience but central to present patterns of urban social geography:

(1) *High-density residential development* High densities of population, combined with high levels of urbanisation and the long history of urban development which preceded the automobile era have made urban land expensive and encouraged the construction of apartment houses (Lichtenberger 1976; Rugg 1979; Sommers 1983). Originating in Italian Renaissance cities, apartment houses, mainly in the form of tenements, had spread through most of Western Europe by the end of the nine-teenth century in a very complicated diffusion process. As Berry (1973) observes, Naples stands as a prototype, with its inner-city neighbour-hoods still dominated by huge blocks of tenements. During the second half of the eighteenth century the apartment house spread to medium-sized cities throughout the continent and in Scotland, often involving the conversion of craftsmen's townhouses into tiny flats.

> These older tenement houses at first supplied the demands of the middle-income group. The barracks for tenants in the lowest social strata were the offsprings of the industrial revolution, created by the laissez-faire housing market . . . [in which] tenement house property became an attractive capital investment for wealthy people. (Berry 1973: 126)

Only in England and Wales, where terrace-building was the high-density urban form, did apartment houses fail to dominate. Moreover, the tradition of apartment-house living has persisted over the postwar period in spite of the residential decentralisation precipitated by the automobile. Indeed, the apartment house has even strengthened its position in England and Wales and parts of northwestern Europe where the tradition of single-family homes has been strongest. The reasons are several. In the first instance, the pressure of population and land values in urban core areas has been compounded by a rapid rise in the price of construction and building materials relative to average wages. As a result, buying a single-family home in the suburbs has gradually been put beyond the reach of a substantial proportion of the population. Secondly, there has been substantial progress in building technology, making possible new forms of apartment houses which have been highly profitable for builders and developers. The expansion of the public housing sector over the postwar period (Fuerst 1974), coupled with the founding of publicly subsidised new towns and satellite towns (Berry 1973) has also exploited these new technologies, creating *grands ensembles* which often consist of 1,000 or more apartments at densities of around 250 inhabitants per ha.

(2) *Neighbourhood stability* In general, Western European cities are characterised by relatively low levels of residential mobility. Combined with the prevalence of brick and stone in housing construction and the logic of good maintenance in expensive housing markets, this means that the life-cycle of urban neighbourhoods tends to be extended. One result of this is that large tracts of European cities provide relatively stable socio-economic environments: something which, in turn, inevitably affects both the nature and the intensity of social interaction and community conflict. The corollary of this overall stability, however, is that social, economic and demographic change tends to be localised, either in volatile areas of inner-city decay or renewal or in newly created suburban or ex-urban environments.

(3) *The scars of war* Europe's history of international conflict is etched on its cities in a number of ways. Defensive sites – particularly hilltop sites – have imposed morphological controls on many major cities (e.g. Edinburgh, Toledo), some of which have never been directly involved in military action. Elsewhere, the changing requirements of fortification have established templates for urban development, with successive sets of city walls not only acting as 'fixation lines' for urban growth but also governing the lineaments of subsequent redevelopment, as in the case of Köln and Portsmouth (Dickinson 1951; Smailes 1966). More recently, the exigencies of the Second World War destroyed much of the fabric of many cities. In some German cities – Koblenz is a good example – between two-thirds and three-quarters of the housing stock was destroyed, while many more of the towns and cities of Britain, France and the Low Countries lost up to one-third of their housing. Even cities which escaped direct bombardment were affected indirectly by the war,

as house-building and infrastructural development came almost to a halt for five years, making for a massive backlog which lasted until the early 1960s. Only the towns and cities of the 'neutrals' – Ireland, Portugal, Spain, Sweden and Switzerland – remained relatively unaffected by the Second World War.

(4) *Municipal socialism* Along with the social, economic and morphological changes associated with the urban transition of the nineteenth century were marked changes in the management of cities. Fear of 'the mob' after the revolutionary movements of the 1840s, combined with the threats of fire and disease, precipitated widespread liberal reform, not only in the structure of urban government but also in legislation relating to every aspect of land use and environmental quality. Soon, municipal governments found themselves presiding directly over a wide range of services in response to the demands of both workers and the new industrial urban elite who wanted better sanitation, decent low-rent housing, cheap transport and better schools. 'The municipality', observed Berry (1973: 124), 'often became the biggest entrepreneur in town'. As the principle of municipal socialism became established and the power of municipal governments expanded, so the *range* of services expanded. In the immediate aftermath of the Second World War, two aspects of municipal socialism emerged as dominant forces in urban development. The first of these was the great expansion of public housing programmes; the second was the introduction of professionalised town planning. Together, they have had a profound influence on the changing social geography of European cities, affecting everything from urban morphology and socio-spatial structure to the quality and location of neighbourhood amenities (Knox 1982b).

Dimensions of Contemporary Urban Residential Differentiation

Ecological Structure

At a general level, the residential structure of the cities of Western Europe is dominated by the same dimensions of socio-spatial organisation that are found in North America and Australasia – socio-economic status and demographic composition. Burtenshaw *et al.* (1981), reviewing the evidence of existing empirical studies, observed that each of these dimensions accounts for between one-quarter and one-third of the total variability in residential differentiation in most cities. Ethnic segregation, classically the third-ranking dimension of residential differentiation is, however, only significant in cities where there has been a large influx of foreigners over the postwar period. It is thus an important component of the ecological structure of Birmingham, Vienna and Zürich, but not of Edinburgh or Barcelona. On the other hand, the interplay of historical and political factors has tended to produce significant dimensions of residential differentiation which do not occur outside Western European cities. In particular, the legacy of nineteenth-century, working-class housing and the more recent acquisition of tracts of public housing – both of which tend to have distinctive demographic profiles – tend to recur in statistical analyses of urban ecology.

In keeping with the classic models of residential structure, socio-economic status tends to be expressed in terms of sectoral differentiation, while demographic differentiation is zonal, and ethnic differentiation – where it occurs – results in localised clusters within the inner city. The spatial expression of those dimensions of urban structure associated with nineteenth-century housing and with public housing, on the other hand, tends to be more variable, since both are highly dependent on the strategies of local bureaucracies: planning in one case, housing administration on the other. At the same time, spatial patterns have been conditioned, in many cities, by the legacy of past phases of morphogenesis. Thus, where urban development has involved discrete growth phases marked by fixation lines such as defensive walls, zonal patterns have been reinforced. Where urban development has followed the establishment of radial routeways, sectoral patterns have been encouraged.

There are other ways in which European cities tend to depart from the classical models of residential structure. One of the most striking is the presence of a stable, high-status area close to the city centre, a feature which is the result of the development of highly prestigious neighbourhoods of a reasonable size before the onset of the urban transition in the nineteenth century. Lichtenberger (1972) and Rugg (1979) also note that the inner city in general tends to contain a variety of land uses, thus making for a less pronounced zone-in-transition. Finally, because the suburbanising influence of the automobile came late in the development of most European cities, their downtown areas have been able to retain their focal position in terms of retailing and social life. At heart, European cities are still pedestrian cities, despite the current trends towards decentralisation, suburban superstores and industrial estates.

In spite of these generalisations, it would be unwise to argue for the idea of a 'typical' West European city. Although there are some shared features in the models which have been advanced for cities in Britain

(Mann 1965; Robson 1975), Germany (Boustedt 1975; Nellner 1976; Riquet 1978), the Netherlands (Buursink 1977), and Spain (Moreno and Miguel 1978), the diversity of urban development within the broader realm of Western Europe defeats the logic of an all-purpose model. There are, however, a few type-examples which between them are fairly representative of European urban structure.

The Continental City: Vienna. The classic 'continental' city can be exemplified by Vienna. Thanks to the work of Bobek and Lichtenberger (Bobek and Lichtenberger 1966; Lichtenberger 1977) it is one of the most clearly understood of all European cities. Their work emphasises the interaction of successive socio-political orders (medieval, absolutist dukedom, high absolutism, nineteenth-century liberalism, modern welfare state) with economic change and technological innovation in transforming land uses, building types and residential patterns as Vienna moved from burger town to *Residenzstadt* and on to manufacturing town, industrial city and, finally, stagnating metropolis. The present-day ecology of the city (Fig. 6.2) is a graphic reflection of these forces. The central commercial area now encompasses the site of an early Roman town (Vindobona) and the *Altstadt*, the medieval core which developed around the castle, cathedral, guildhall and marketplaces which served as focal point for medieval society. While these early features remain, together with much of the medieval street plan, much of this central area is now dominated by the institutional architecture of the Renaissance and baroque periods, when the Hapsburgs indulged in Grand Design in order to demonstrate imperial power and glory. This ideology overlapped into the urban transition of the industrial period, when Emperor Franz Josef encircled the Altstadt with the *Ring*, a series of boulevards flanked by public buildings and parks, laid out in 1857 on the site of the old town walls. Within a short time, however, the city came to be dominated by the suburbs and industrial sectors beyond the Ringstrasse. Large factories and their associated tenement housing were built along the axes formed by the Danube and by the railways which radiated to the further points of the Empire (Mayer 1978). Commercial areas emerged along major roads and tramways and in the older towns and villages absorbed by the expanding city, producing a mix of sectoral and zonal differentiation (Figure 6.2) which persisted as the ecological framework of the city during the postwar period when growth petered out, and the town planners of the welfare state stepped in to thin out, tidy up and re-zone the city.

Figure 6.2: The Ecological Structure of Vienna

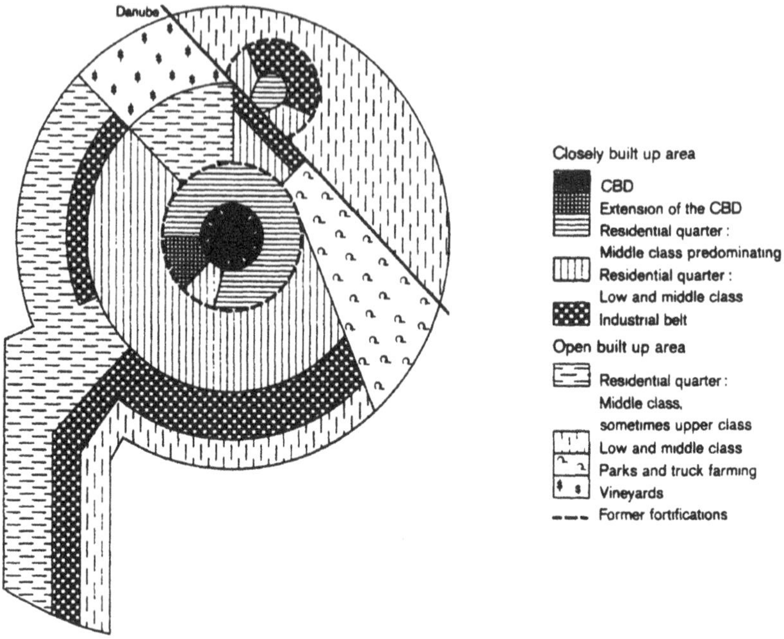

Source: Lichtenberger (1977).

The Southern City: Barcelona. The cities of southern Europe present a more confused amalgam of traditional and modern structures. Barcelona provides a good example, although at first sight it does appear to reflect a relatively simple chronological division between a medieval core, a surrounding new town and sprawling peripheral suburbs (Figure 6.3). In detail, however, the city reflects a complex succession of interactions between social and economic organisation and political ideology. The medieval core of the city, or *casco*, was developed during the thirteenth and fourteenth centuries, providing a template which remained largely unchanged until the beginning of the eighteenth century, when the early Bourbons, like the Hapsburgs in Vienna, began to reconstruct parts of the city, building a citadel in the Grand Manner. Meanwhile, the military authorities, concerned about sanitation, and worried about security in a notably rebellious part of Spain, opened, widened and straightened many of the streets (Wynn 1979). Eventually, in 1854, the military consented to the development of the *Ensanche* area beyond the medieval walls. The basis of this development was a plan drawn up by Ildefonso Cerda, a pioneer in urban planning who envisaged a system of neighbourhoods based on the *manzana*, a square block lined by apartment houses on two sides. In the spirit of nineteenth-century capitalism, however, the plan was only loosely adhered to. While the street layout was largely respected (Figure 6.3), little else was.

> The *manzana* was built up on all four sides . . . and within. Parks and gardens were encroached upon or disappeared altogether; manzanas destined for schools, markets and social centres in the Plan Cerda were used for house construction and commercial and industrial buildings. Instead of the egalitarian polycentricism foreseen by Cerda, middle-class residences tended to be built around the central commercial axis . . . with the working classes resident in the dilapidated houses of the Old City and in the sub-equipped, poorly connected periphery in and around the old nuclei which included the main concentrations of industry, forming mixed residential/industrial zones. (Wynn 1979: 187-8)

Figure 6.3: The Spatial Structure of Barcelona in the mid-1970s

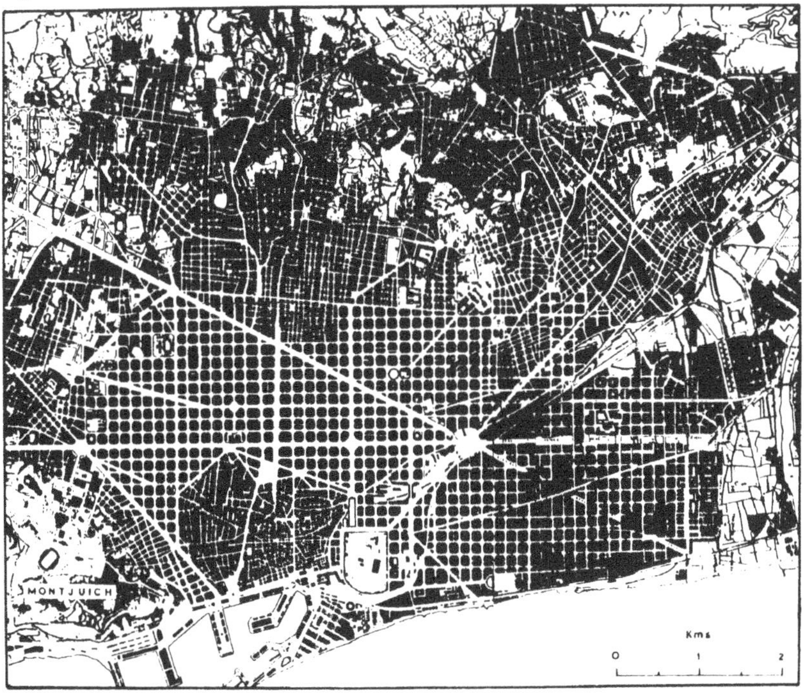

Source: Naylon (1981).

During the early twentieth century the more dynamic elements of the Catalan middle classes attempted to create an 'Imperial Barcelona', while the advent of the Second Republic in 1931 saw the beginnings of radical reform in the face of rapid growth and extreme housing shortages. The establishment of the Franco regime in 1939, however, put an end to all progressivism (Naylon 1982). Because of bureaucratic centralisation, municipal fragmentation and financial dependency under Franco, the postwar development of the city has been haphazard and unconstrained:

A weak official planning policy, coinciding with enormous population pressure ... and a corresponding demand for building land and housing, could not help but produce a speculative fever characterised by rocketing land prices, building anarchy and galloping deficits in infrastructures and services. The aggressive behaviour of landowners, development companies, building firms, banks and finance houses, the *Cambra de la Proprietat* (Chamber of Property) and other private interests bore upon town councils from the very beginning and accumulated wealth in favoured hands while virtually pauperizing large sections of the population. (Naylon 1982: 23-4)

The contemporary social ecology of the city, while set in the superficially straightforward framework depicted in Figure 6.3, is thus somewhat chaotic, with continued inmigration not only exacerbating the squalor of inner-city districts, but also spilling over into extensive peripheral tracts of high-density apartment housing, and sustaining a residual but substantial squatter population in the *barracas* which occupy the fringes and interstices of the city (Lowder 1981).

The Industrial City: Birmingham. Among the industrial cities of north-western Europe, Birmingham provides an outstanding example. Here, the absence of any previous urban development of any significance, combined with extensive heathland environs, provided an uncluttered stage: the city's present structure (Figure 6.4) is a product of nineteenth-century urbanisation, and twentieth-century suburbanisation and renewal (Wise and Thorpe 1970). The original settlement, not much more than an industrial village, grew without a charter, and its crafts were therefore unshackled by the restrictions of guilds. Having specialised in metal manufacture, it was well placed for expansion during the industrial revolution. By 1838, when it was made a borough, its population had grown to 186,000, and its spatial organisation was

Figure 6.4: The Ecological Structure of Birmingham

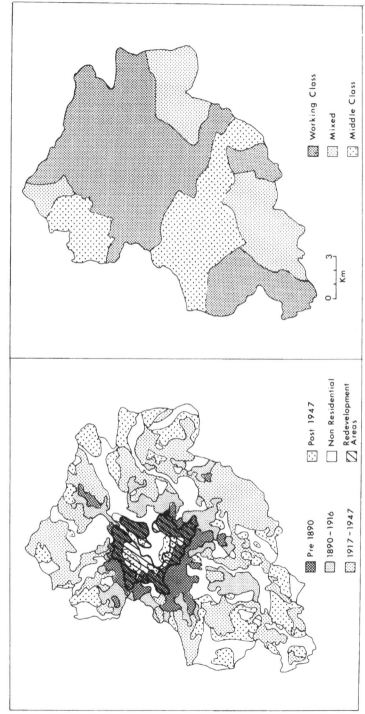

Source: Adapted from Department of the Environment (1977).

dominated by the functional specialisation of particular trades. It was not until the last third of the nineteenth century, however, that Birmingham matured into a major industrial city. Backed by what had become a well-developed canal system and stimulated by the opening of the London-Birmingham railway, large-scale factory units began to appear as the advantages of agglomeration, localisation and scale began to tell. With plenty of cheap land available, the main thrust of development took place on the outskirts of the existing area, with the Birmingham Small Arms factory at Small Heath, Cadbury's new site at Bournville, the Austin motor works at Longbridge, and the General Electric plant at Witton acting as nuclei for the spread of the city.

Meanwhile, in response to the slums and insanitary conditions brought by such rapid urbanisation, liberal reform began to have an impact on the form and structure of the city. Led by Joseph Chamberlain, Birmingham progressed towards being the 'best-governed city in the world'. The early imprint of civic pride and municipal socialism was expressed in terms of the erection of public buildings, widening of streets, slum clearance and the provision of public utilities, but it was not long before the city council came to exert a more strategic influence on urban development through the public transport system and through housing legislation. As this influence began to take effect, however, the advent of the automobile triggered the modern phase of decentralisation, creating extensive tracts of suburban development on the margins of the city, in ribbons along the major roads radiating from the city, and in clusters around nearby villages. After 1947 the rapid ascendancy of professional planners began to constrain this development through zoning and green-belt strategies. At the same time, the structure of the central area was transformed through commercial redevelopment by the city council in partnership with private enterprise. Most important of all, however, was the city council's role as an agent of change in the inner-city neighbourhoods and peripheral suburbs, where large tracts of public housing were erected – much of it in the form of high-rise apartment houses – in an attempt to combat the legacy of substandard housing and housing shortages. By 1981 35 per cent of the city's housing stock was council-owned, representing a major component of the city's ecology and tending to mask the combination of sectoral and zonal differentiation which had been the product of successive phases of growth based on urbanisation in the private arena (Figure 6.4).

Ethnicity and Segregation

Although ethnic segregation represents a minor dimension of residential differentiation in general terms, it must be acknowledged that it is a distinctive feature of a growing number of cities. Indeed, a considerable literature exists on the Asian, West Indian and Irish sub-areas within British cities (Dahya 1974; Hiro 1973; Jones 1979; Lee 1977); and there is a growing literature on the emerging social geography of ethnic minority groups — mainly guestworkers — in the cities of central and northwestern Europe (Eggeling 1981; Jones 1983; Lichtenberger 1982; Mik 1983). It should be stressed at the outset, however, that this ethnic segregation does not bear close comparison either in its nature or its intensity with the ethnic segregation of North American cities. Thus, while the spatial configurations are the same — inner-city clusters — the overall degree of geographical segregation is somewhat lower and the processes involved are somewhat different, with relatively less emphasis being given by most analysts to the role of discrimination and more to the role of the 'fabric' effects of housing and labour markets and to the internal cohesiveness of the ethnic minorities involved. In Britain most larger cities contain sizeable Asian and West Indian communities, together, in some cases, with an Irish community. Other minority groups, such as the Maltese in London (Dench 1975), are not large enough to have a significant impact. Birmingham again provides a good illustration of the geography of the major groups. As Jones (1979) has shown, the ethnic sub-areas of Birmingham are not exclusively dominated by one group but are rather mixed neighbourhoods, with the non-white population amounting to only 15 or, at most, 20 per cent of the total in each cluster. There is evidence, however, that there is a significant degree of segregation *within* these clusters, with local concentrations of Indians, Pakistanis, Bangladeshis, Jamaicans, Trinidadians, and so on. Moreover, these clusters seem to be intensifying, despite the increasing role, noted by Peach (1975), of public housing in the geographical assimilation of West Indians. Jones (1979) also draws attention to the location of the ethnic clusters: in a zone just outside the inner city proper. This is a consequence partly of structural, partly of institutional factors. The innermost neighbourhoods of Birmingham have proved unsuitable to immigrants and their families because the small, nineteenth-century terrace homes are unsuitable for larger, extended household units; and in any case a good deal of territory in these neighbourhoods is municipally owned and scheduled for comprehensive redevelopment. The immigrant clusters have, therefore,

Figure 6.5: The Distribution of Turkish and Yugoslav Guestworkers in Vienna

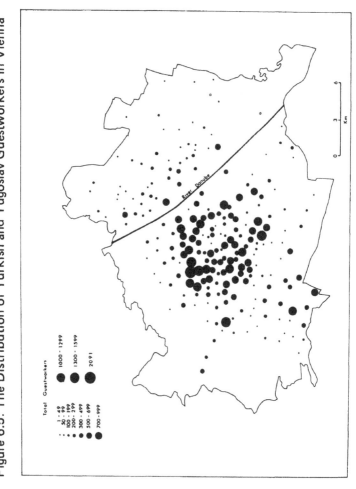

Source: Leitner (1981).

been pushed out and have taken over neighbourhoods of larger Victorian and Edwardian housing which has filtered down the social scale, and which is not particularly attractive to the indigenous population because of the lack of amenities such as garage space. The net result is that the immigrant clusters were 'not typically sited in the slum-ridden heart of the city, but in the tree-lined, often attractive townscapes of the middle ring' (Jones 1967: 22). Once established in these areas, immigrant clusters have tended to persist, partly because of ethnic cohesiveness (Boal 1976) and partly because of the truncation of inmigration which might otherwise have fuelled a process of invasion and succession.

Clusters of migrant workers in the cities of continental Europe are a more recent phenomenon, but already they exhibit a reasonably consistent pattern. The geography of migrant workers is such that ethnic minority populations are dominated by different groups in different cities – Yugoslavs and Turks in Duisburg, Frankfurt, Köln, and Vienna; Algerians, Italians and Tunisians in Paris; and Surinamese and Turks in Rotterdam, for example – yet their spatial behaviour *within* these cities seems to be fairly straightforward. Just as the migrants are replacing the lower echelons of the indigenous population in the labour market, so they are acting as a partial replacement for the rapidly declining indigenous population in the older neighbourhoods of privately rented housing near to the sources of service employment and factory jobs (Solomos 1982). This produces the kind of spatial pattern depicted in Figure 6.5 for Yugoslavs and Turks in Vienna, with marked concentrations throughout the inner city and in outlying neighbourhoods adjacent to industrial districts. Quite simply, the overriding priority for migrant workers is to live close to their jobs in cheap accommodation. This applies to all national groups, so that the same spatial pattern has persisted even as the culturally more alien Turkish, Yugoslavian and North African populations have replaced older and more familiar groups such as Greeks and Spaniards.

Once again, however, it is clear that there is a high degree of segregation *within* immigrant areas. As Clark (1975) and O'Loughlin (1980b) have demonstrated, individual apartment buildings are predominantly – sometimes exclusively – inhabited by a single group. This occurs because knowledge of vacant and available dwellings is passed from one immigrant to fellow countrymen. At the broader, neighbourhood, level, segregation was fuelled, in the first place, by chain migration, with extended families and friends from rural villages following one another to the same neighbourhood. This was reinforced by the tendency of some employers to recruit from particular regions. Since the recruitment bans imposed by most countries after 1973, segregation has been maintained through a combination of ethnic cohesiveness and fabric effects, though there is some evidence of spatial adjustment as a product of the differential demographic and employment structures of different national groups (Döpp and Leib 1980; Jones 1983; Mik 1983).

Urban Problems and Planning Responses

Ethnic segregation is, of course, closely associated with syndromes of urban deprivation and inner-city decline. But ethnic sub-groups — like the elderly, the homeless and the poor in general — are clearly the victims rather than the causes of urban problems. It is the changing economic, demographic and morphological structure of cities which is the chief source of urban problems in Western Europe, as elsewhere. Naturally, the exact *nature* of urban problems will depend on the particular stage of urban development and the consequent nature of economic, demographic and morphological change. Nevertheless, European urbanisation has followed a sufficiently uniform path to have precipitated a number of common problems. These cannot be dealt with in detail here, but it is important at least to establish their principal dimensions and their immediate context.

Urban Change and Social Problems

Given the history of European urbanisation, it is not surprising that one of the most general and most pressing urban problems concerns the age and conditions of the housing stock. As Table 6.4 shows, the majority of dwellings in many cities date from before the Second World War, and it is quite common for between a quarter and half of the housing stock to lack what have come to be considered as 'basic' amenities: an inside toilet, a bath or shower, and central heating. Moreover, the nature of urban development has of course meant that the bulk of these dwellings are localised in inner-city neighbourhoods while the newer, better-equipped housing — both private and public — is found in peripheral suburbs, along with newer factories, shops, offices and infrastructural components. This decentralisation has itself been the source of serious problems. Because the decentralisation of housing, in general, preceded the decentralisation of employment, there occurred an increasing mismatch between residential location and workplace which aggravated the problems of urban transportation faced by cities whose morphology is geared more to pedestrians and streetcars than to automobiles.

Decentralisation has also heightened conflict over land use at the urban fringe and, as urban decentralisation has outpaced the reform of political boundaries, it has generated the kind of fiscal problems associated with the fragmented governments of North American cities (Centre for Advanced Land Use Studies 1981). More specifically, decentralisation has added a new dimension to the problem of service delivery, with the demands of large suburban populations conflicting with the economic and administrative logic of centralisation. In the faster-growing cities of southern Europe, the problem of service provision is often absolute rather than relative. In Barcelona, for example, there was a deficit of 100,000 hospital beds and nearly 180,000 school places in 1970, and in some peripheral neighbourhoods even basic services such as electricity supplies, water, drainage, road surfacing and flood control measures have had to be organised and paid for by residents themselves as the city government has been unable to cope with urban growth (Naylon 1982; see also Williams 1981).

Table 6.4: Housing in European Cities, *c.*1975

	% of total dwellings built		% of dwellings with a w.c., a bath or shower and central heating
	1946-60	After 1960	
Arhus	23.5	30.0	97.2
Basel	14.0	22.0	72.7
Bergen	23.2	18.5	69.2
Bologna	37.3	26.4	n.a.
Bordeaux	13.0	16.3	n.a.
Caen	41.4	23.0	n.a.
Düsseldorf	43.0	17.7	41.0
Frankfurt am Main	35.4	16.5	39.0
Gothenburg	19.7	29.6	76.4
Grenoble	27.3	29.3	n.a.
Hamburg	37.2	20.9	47.0
Helsinki	27.8	33.7	74.3
Köln	39.0	21.1	32.6
Le Havre	31.8	13.6	n.a.
Lyon	17.5	18.6	n.a.
Milan	32.6	25.9	n.a.
Naples	25.3	27.5	n.a.
Oslo	28.8	17.5	72.6
Rome	35.9	36.7	n.a.
Rotterdam	18.0	17.1	n.a.
St Etienne	16.0	13.0	n.a.
Stockholm	18.6	14.4	82.5
Utrecht	16.1	25.5	27.3
Zürich	26.0	12.6	86.7

Source of data: United Nations (1980).

Meanwhile, problems associated with the physical deterioration of housing in older, inner-city neighbourhoods have been compounded by related social and economic problems. As employment has decentralised, inner-city areas have become increasingly characterised by unskilled and semi-skilled workers with high rates of unemployment and poor prospects of improvement; and by a residual population of vulnerable and indigent households, including many elderly persons (Pierce and Hagstrom 1981). This, in turn, reflects the influence of demographic trends. The downturn in the birth-rate which began around 1964 in northwestern Europe precipitated rapid changes in household composition, the consequent ageing of local populations being most pronounced in inner-city areas vacated by younger cohorts (Burtenshaw *et al.* 1981). This localisation of poor and vulnerable households in deteriorating physical environments has created a vicious cycle of poverty in which a wide spectrum of social problems are generated, sustained and amplified. At the same time, it must be recognised that the creation of 'vulnerable' or 'problem' groups in the first instance is a product of the social polarisation stemming from economic organisation rather than from spatial organisation or environmental circumstances.

What does all this add up to in terms of the overall pattern of urban problems? Recent research by O'Loughlin (1983) suggests that, although the *degree* of socio-spatial inequality in West German cities broadly parallels that of North American cities, the *localisation* of disadvantages in German inner-city areas tends to be rather less pronounced. We can probably extend this comparison to include the cities of most of the rest of Western Europe, although there has been very little comparative research in this area. From the evidence which is available, it has been argued that the spatial distribution of urban problems in European cities tends to be broadly accumulative, with an overlap in the distribution of deprivation resulting in pockets of 'multiple deprivation' (see, for example, Aase and Dale 1978; Bentham and Moseley 1980; Millar 1980). This does not mean that most neighbourhoods do not experience significant problems; research has shown that problems associated with a wide range of life domains – including health, housing, employment, education, transport and leisure – are experienced to some degree by residents of most neighbourhoods (see, for example MacLaran 1981). Indeed, some studies (Coulter 1978; Knox and MacLaran 1978; Peters 1979) have pointed to the existence of different *syndromes* of socio-economic problems associated with different sets of neighbourhoods. Similarly, Madge and Willmott (1981) have shown that both the pattern and the intensity of socio-economic problems can vary significantly between comparable inner-city neighbourhoods in different cities – London (Stockwell) and Paris (Folie-Méricourt). In Stockwell, for example, household disadvantage was, comparatively, most pronounced in terms of low incomes; while in Folie-Méricourt it was most pronounced in terms of overcrowding and lack of housing amenities. Moreover, whereas there were two broadly separate syndromes of disadvantage in Stockwell (one related to housing conditions, the other to income, health and leisure), there was no distinct pattern of disadvantage in Folie-Méricourt. Thus, although multiple deprivation in inner-city areas may be the most *salient* social problem for many cities, it is by no means simple or uniform in its manifestation.

Town Planning and Urban Policies

As Berry (1973) has emphasised, town planning has not only been a major product of European urbanisation; it has also been a major factor in shaping and modifying the character of European cities themselves. Burtenshaw *et al*. (1981) recognise five major 'traditions' of European urban planning, each of which has contributed a distinctive legacy to different groups of cities at different points in time.

The Authoritarian Tradition. This is the largest tradition, and its legacy can be seen in a variety of forms. Its high point came in the Renaissance when, as we have seen, military authoritarianism was reflected in new settlements like Palma Nova. Subsequently, the stamp of authoritarian planning flourished during the despotic regimes of the baroque, producing extravagant symbolisations of wealth, power and destiny. During the nineteenth century, when utilitarianism and *laissez-faire* attitudes prevailed, the authoritarian tradition was muted, though its imprint was maintained in larger cities. The best example is the work of Baron Haussmann in Paris on behalf of the third empire. Here, the motivation was a mixture of the vanity of the elite, the need to create employment, and the urge to control 'mob' behaviour by making streets more difficult to barricade. Haussmann cut new boulevards across the plan of medieval Paris, lining them with trees and creating over 20 urban parks where there had previously been none. Meanwhile, his example inspired similar schemes in other French cities, and Franz Josef's development of the Ringstrasse in Vienna and Cerda's plan for the expansion of Barcelona were also influenced by the 'Haussmannisation' of Paris.

During the twentieth century the authoritarian tradition has been best represented, not surprisingly, in the fascist regimes of Italy, Spain and Germany. Hughes (1980: 427) has described Mussolini's attempt to create 'La Terza Roma' as a style which had 'a jackboot in either camp: one in the vision of Ancient Rome and the other in the vision of a technocratic future'. Enraptured by modernisation, technology and the cult of masculinity, Mussolini adopted Marinetti's Futurism as the official style for Italian Fascism, somewhat in contradiction to the strong element of anti-urbanism embodied in fascist philosophy (Treves 1981). In Spain the authoritarian tradition was exemplified by the 1941 Plan for Madrid (Wynn and Smith 1978), while the development of anti-Franco Barcelona, as we have seen, was deliberately neglected. In Germany, where party ideologists condemned the whole modern movement in architecture as alien and culturally degenerate, the most striking legacy of authoritarianism is, ironically, to be found in the officially encouraged domestic architecture of the period: variants of the Hansel and Gretel cottage and of Tyrolean designs which were intended to reinforce the idea of a distinctive German culture rooted in the virtues of rural peasantry (Knox 1982c; Mullin 1981, 1982). Nevertheless, the Third Reich did allow itself to indulge in showpiece ceremonial architecture such as the House of German Art in Munich and the Zeppelin Field and Congress Hall in Nuremburg (Gloag 1979). Since 1945 the authoritarian tradition has had little direct influence in Western Europe, though it has been argued that it can be seen in the strong centralism of French planning, particularly under de Gaulle (Burtenshaw *et al.* 1981).

The Romantic Tradition. This is a tradition which dates from the late nineteenth century, representing a reaction to the harsh environments of industrial cities and to the 'Haussmannisation' of 'traditional' urban environments in Europe's capital cities. The inspiration of the movement was Camillo Sitte, who emphasised the role of town planning as a creative art and called for irregularity and imagination in place of abstract, geometrical and monumental planning. His ideas were utilised directly in plans for Salzburg and Vienna, and they can be traced in particular stages of the development of many other cities, including Amsterdam, Brussels, Gothenberg, Madrid, Mainz, Munich and Wiesbaden.

The Tradition of Organic Planning. This is another product of nineteenth-century European urbanisation. Its most articulate representative was Patrick Geddes, whose reaction to the urban problems of the late nineteenth century was based on the idea of understanding the city as an organism whose growth had to be properly guided if its worst tendencies were to be avoided. The organic tradition brought a more 'scientific' and 'rational' view to bear on town planning, aiming at organic growth within a continuously monitored framework. The classic example of this tradition is Abercrombie's 1945 plan for Greater London in which green-belt and overspill strategies were first systematically developed. The 1970 Strategic Plan for the South-East is a direct descendant; and the organic approach can also be recognised in plans for Hannover, Manchester, Stockholm and Randstad.

The Utopian Tradition. This was another response to the social and environmental conditions of the industrial revolution, and it can be traced to attempts by Owen (in Scotland) and Fourier (in France) to create model communities in the early nineteenth century. By mid-century the idea had been taken up by a number of industrialists, mainly for the benefit of their own workers and their families. Hence Titus Salt's Saltaire, Lever's Port Sunlight, Cadbury's Bournville and Rowntree's New Earswick, together with several ventures in the Ruhr by Krupp. During the later part of the nineteenth century the utopian idea was sustained by the growing influence of intellectuals like Disraeli, Morris, Ruskin and Zola, but the only significant manifestation of utopianism was in 'five per cent philanthropy': the creation of sanitary but cheap accommodation for the working classes by philanthropists willing to accept modest levels of profit for investments in a good cause (Tarn 1973). It was when Ebenezer Howard systematically organised the ideas of the utopian movement in 1902 that the full expression of the tradition emerged in the shape of garden cities and suburbs. Howard himself was closely involved in the Garden City Pioneer Co. Ltd, which promoted the development of Letchworth, the first full garden city. The success of Letchworth and, soon afterwards, Welwyn Garden City and Hampstead Garden Suburb, inspired a rash of similar developments all over Europe: Margarethenhöhe and Romerstadt (Germany), Floréal (Belgium), Tiepolo and Campo dei Fiori (Italy) and Hirzbrunnen (Switzerland), for example. In the postwar period the garden-city idea has given way to the new town, in which utopian ideals have been displaced to a considerable extent by strategic ideas ranging from metropolitan overspill to growth-points in areas of economic decline. Nevertheless, the utopian tradition remains an overt influence on all of the new towns of Western Europe. There are now more than 30 new towns in the United Kingdom, where they have become an important and distinctive component of postwar urbanisation (Bourne 1975). On the continent new towns have mainly been developed as extensive satellites to major cities, and they are most numerous in France, West Germany, the Low Countries and Scandinavia (von Einem 1982; Goldfield 1982; Tuppen 1983).

The Technocratic Tradition. The idea that buildings and cities could be designed and run like machines forms the basis of this tradition. Its origins lie in the artistic and intellectual reaction to the romanticism of the Arts and Crafts movement, Impressionist painting and utopian planning of the nineteenth century. This reaction, inspired by Cubism, sought to dramatise modern technology, using an anonymous and collective method of design in an attempt to get away from 'capitalist' canons of taste. Thus emerged the Constructivist and Futurist movements, the Bauhaus school and, later, Les Congrès Internationaux d'Architecture Moderne (CIAM) and the Modern Architecture Research Group (MARS), who believed 'that their new architecture and their new concepts of urban planning were expressing not just a new aesthetic image but the very substance of new social conditions which they were helping to create' (Carter 1979: 324). While the 'White Gods' of the tradition were Walter Gropius and Mies van der Rohe of the Bauhaus it was Le Corbusier – 'Mr Purism' (Wolfe 1981) – a Paris-based Swiss, who provided the focus and the inspiration for technocratic town planning. His houses were 'machines for living', and his idea of urban design was based on high-density, high-rise apartment blocks, elevated on stilts and segregated from industrial districts and transport routes by broad expanses of public open space: a functional, 'non-bourgeois' city. In the first instance, examples of technocratic design and planning were limited to small projects, mainly of worker-housing sponsored by socialist governments. After the Second World War, though, technocratic design became pervasive, part of the 'International Style' which was avant garde yet respectable and, above all, cheap to build. It has been the technocratic tradition which has, more than anything else, imposed a measure of uniformity on the heterogeneity of European cities.

Postwar Orientations. It should be stressed that these traditions have influenced the character of individual cities in a piecemeal and complex fashion, as Costa (1977) and Fried (1973) have demonstrated in the case of Rome. Moreover, the professionalised town planning of the postwar period must be seen as an amalgam of traditions, including some (spatial determinism and the systems approach) derived from

North America; and its impact on the cities of Western Europe can best be understood not so much in terms of particular ideologies as in terms of particular policy spheres. In this context, an excellent review is provided by Burtenshaw *et al.* (1981), while useful reviews of planning policy in individual countries are provided by Albers (1980), Arcangeli (1982), Blijstra (1981), von Einem (1982), Goldfield (1982), Hajdu (1983); Short (1982), and Wynn (1983).

The point which should be made here is that, among the diversity of individual national circumstances and responses, and in addition to some of the almost universal components of contemporary town planning (e.g. urban renewal schemes, housing rehabilitation programmes, pedestrianisation schemes, green belts, new towns), there are certain themes which are distinctive to Western European urbanism. One of these concerns the chronology and orientation of urban planning in the postwar period. Three broad phases can be identified. The first was that of the immediate postwar period, when the focus of attention was literally on cleaning up the aftermath of war and making good the backlog of housing and basic amenities. Because of the sheer pressure of these basic needs, few cities were able to be innovative: the chance to redesign urban morphology had to be sacrificed to speed and economy. By the late 1950s the combination of increasing affluence and population growth, together with the prevailing of planning theory, initiated the second phase, one of centralised, technological planning whose object was to establish and maintain specific criteria and standards for urban development. Above all, the aim was to make cities 'modern' and 'efficient'. This was the phase of relentless slum-clearance projects, city-centre redevelopment schemes, urban motorways, and massive public housing projects: 'heroic' planning. The third phase, beginning around the critical watershed of the early 1970s, has been one of reaction and reappraisal. The mistakes and excesses of the second phase have introduced new objectives such as equality and public participation, while changes in economic and demographic circumstances have modified the whole tenor of urban planning. Renewal and redevelopment have been superseded by rehabilitation and regeneration.

Throughout these phases the legacy of prewar European urbanisation has also led to some distinctive and broadly shared planning responses. One of the best examples is in terms of transport planning, where the legacy of high-density residential development and the prevalence of apartment-house dwellings and the long-standing preference (or acceptance) of public transport (McKay 1976) has encouraged the development of sophisticated rapid-transit systems. Since the 1960s a number of cities (including Barcelona, Frankfurt, Hamburg, Köln, London, Madrid, Milan, Munich, Newcastle, Paris, Rome, Stockholm and Vienna) have built or extended subway or metro systems. As a result,

> whereas in North American metropolitan areas many traditional city centres have begun to rot away, throughout Europe public and private development are combining to preserve, rehabilitate and/or construct new city centres that continue to dominate urban life. (Berry, 1973: 144)

Another example is the concern with civic design and conservation in European cities which clearly stems from the rich legacy of pre-modern urbanisation. As Hall (1969) observes, there is widespread feeling that the city is 'the custodian of Europe's cultural patrimony'. Consequently, grass-roots conservation movements have developed in many cities, and conservation planning has become an integral part of the formal planning process throughout most of Western Europe (Appleyard 1979; Kain 1980). Moreover, both the Council of Europe and the European Community have become heavily committed to urban conservation, not only on cultural grounds but also because of the economic logic of conserving environments which are attractive to tourists.

7 SOCIAL WELL-BEING IN EUROPE'S WELFARE STATES

Western Europe is a comparatively prosperous region. Growth in West European *per capita* incomes averaged more than 4 per cent per annum during the first two decades of the postwar period, compared to the rate of around 1 per cent per annum which had characterised the entire period of industrial development since 1800. Even during the periods of recession and inflation in the 1970s, most countries' growth-rates remained at respectable levels by historical standards (Aldcroft 1978; Lawson and George 1980). Europe's inhabitants, on average, now consume more than twice the quantity of goods and commercial services that they did in 1950. Purchasing power has risen everywhere to the extent that basic items of food and clothing now account for less than half of household expenditure (in most of Western Europe the figure is in fact closer to 30 per cent), leaving more resources for leisure and consumer durables (OECD 1980). Nearly nine out of ten households have a refrigerator, nearly half have a television set, and over two-thirds have a car (Euromonitor 1977). Meanwhile, the development of welfare states has helped to maintain purchasing power during periods of recession and to ensure at least a tolerable level of living for most groups at all times. Whereas expenditure on defence has remained fairly constant at between 1 and 5 per cent of countries' GNP, expenditure on social welfare has generally increased from between 2 and 6 per cent in the 1950s to between 8 and 16 per cent in the 1970s (Pluta 1978). Overall, *per capita* social security transfers have increased roughly threefold in real terms since 1960 and eightfold since 1950.

Yet, as we have seen, the overall prosperity of the region must be set against the differential effects of economic, political, demographic and social change. Among the overlapping, intersecting and interacting processes which affect social well-being at the regional and local level are population stagnation and selective migration (Chapter 2), political centralisation and regional separatism (Chapter 3), the marginalisation of rural communities (Chapter 4), structural economic change (Chapters 4 and 5), metropolitan decentralisation (Chapter 6), and the increasing involvement of central and local government in all spheres of economic and social life (*passim*). The question is, 'What do they all add up to in terms of who gets what, where?' What, in other words, is the extent of regional inequality in social well-being? And is there any evidence of change in the pattern or intensity of inequality?

The Extent of Regional Inequality

Because of governments' increasing role in managing various aspects of social well-being, data are now available for a wide spectrum of regional socio-economic indicators. In many countries concern over the broader implications of economic change for people's quality of life has prompted the compilation of comprehensive surveys and data banks: some of the best examples are from Scandinavia (Allardt 1981; Hansen and Geckler 1978; Johansson 1973; Ringen 1974; Vogel 1981; Wagtskjold 1982). As a result, it has been possible to gain a detailed impression of both the nature and extent of regional variations in social well-being (see, for example, Asheim 1978; Goodyear and Eastwood 1978; Knox 1974; Knox and Scarth 1977; Walter-Busch 1983). Lack of comparability in national statistical series has inhibited cross-national studies at the regional scale of analysis (though the OECD has an ambitious social-indicator programme (Verwayen 1980), and the European Community publishes some standardised regional statistical series). Nevertheless, it is clear that some of the basic parameters of social well-being exhibit a disturbingly sharp gradient from one part of Western Europe to another. Evidence has already been presented (p. 139) in relation to core-periphery contrasts in economic development, and these are paralleled by patterns of income and consumption. *Per capita* incomes throughout Switzerland, for example, are more than twice those in any Spanish region, while television ownership in all West German regions is double the rate in southern Italy. The most useful single indicator of social well-being, though, is probably the infant mortality rate. Here again we have seen the sharp polarisation which exists: the more prosperous regions of northwestern Europe generally experience only half the rate of mortality of the depressed rural regions of southern Europe (Figure 2.2). It will be noted, though, that the spatial pattern is dominated by a north-south differential, unlike the core-periphery pattern associated with indicators of economic activity, income and consumption. This difference can be largely attributed to the more highly developed welfare states which exist in the northern periphery of Western Europe; and it serves to emphasise the difference between economic health and the broader concept of social well-being (Smith 1977; Knox 1975).

Figure 7.1: Perceptions of Affluence: Percentage of Respondents Claiming a 'fairly high', 'high' or 'very high' Standard of Living

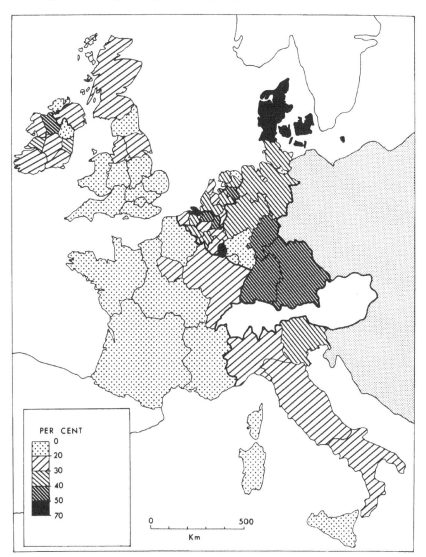

Source of data: Eurobarometer 5 (1976).

It is useful to make a further distinction between these variations in objective circumstances and variations in people's *perceptions* about economic prosperity and social well-being. The Eurobarometer series provides some interesting data here, despite the restrictions of its territorial coverage. Drawing from the 1976 survey (Eurobarometer 5), Morrison (1977) and Van Praag (cited in Dennett, James, Room and Watson 1982) have pointed to the wide differences which exist between countries in the perception of affluence and poverty. Thus, whereas the proportion of Danes and Dutch who consider their income to be less than the necessary minimum is around 20 per cent, the proportion of Irish and Italians who feel similarly disadvantaged is around 40 per cent. At face value, this seems to reflect national variations in objective economic circumstances. Yet only 25 per cent of Europeans with low incomes actually perceived themselves as poor – or were willing to say so – and only 10 per cent of those in the highest income brackets thought of themselves as wealthy. Moreover, the relationships between objective circumstances and subjective evaluations of them are known to be complex, conditioned not only by present circumstances but also by variations in people's past experiences, their expectations and aspirations and their reference groups (Knox and MacLaran 1978; McKennell and Andrews 1980). It is not surprising, therefore, to find that regional variations in perceptions of affluence do not demonstrate any straightforward relationship to the various 'objective' dimensions of regional prosperity that we have examined. Figure 7.1, which is based on the proportion of households in each region who rated themselves favourably (5 or more on a 7-point scale) in terms of their standard of living, suggests that there is a general distinction within the European Community between the pronounced sense of prosperity which characterises Denmark, southern Germany and parts of the Benelux countries, and the sense of relative deprivation which dominates most of France, Ireland, Italy and the United Kingdom: a division which cuts across all of the major socio-economic sub-divisions of European regions.

The Persistence of Inequality

Given these disparities in affluence, social well-being and people's sense of prosperity, the question arises as to their persistence. Changes in regional disparities should first be placed within the context of distributional change at the national level, however. Unfortunately, the evidence is ambiguous. In part, this is because of the complex methodological and analytical problems involved (Atkinson 1975). Some research points towards a convergence of well-being: the work of Pen (1979), for example, on income distribution in the Netherlands. Pen claims that income inequality in the Netherlands halved between 1938 and 1978, partly because of progressive changes in taxation and social security, and partly because of the equalising effects of educational provision. Most analysts have been less optimistic, however. Lawson and George (1980), reviewing postwar change in Belgium, France, Ireland, Italy, West Germany and the United Kingdom, conclude that:

> In most countries there appears to have been a slight redistribution of income away from the very rich, but this seems to have been mainly in favour of the relatively well-off. Moreover, in all the countries studied the richest 5 per cent and 10 per cent have retained between two-and-a-half and three times their 'parity shares', the shares they would have had if incomes had been equally distributed ... As regards the poorest groups, only the German data show a slight and sustained improvement over the whole post-war period; but, even so, the bottom 20 per cent and 30 per cent received a mere one percentage point more of total income in 1973 than in 1950. (p. 235)

Similar conclusions emerge from comparative reviews undertaken by the OECD (1980), by the European Community (Commission of the European Communities 1979) by Jain (1975) and by Kraus (1981). In short, the *status quo* predominates, but it is qualified by some convergence as far as certain groups are concerned.

Relative Change in Regional Inequality

How does this translate into change in relative levels of well-being between different cities and regions? The issue is particularly intriguing in view of the conflicting outcomes associated with different theoretical perspectives on the dynamics of urban and regional change. In overall terms, changes in regional inequality can be categorised into one of three theoretical outcomes: convergence, divergence and the *status quo*. The convergence hypothesis refers to a reduction in the level of differentiation between regions, and it is supported by two major sets of views. First, there is the belief that, in advanced industrial economies, the mobility of labour, together with efficient transportation systems and the rapid diffusion of innovations and information, will lead to the eventual elimination of interregional differences in wage-rates and profit levels, thus making for a relative convergence of levels of social well-being. This, of course, is a fundamental tenet of neo-classical economic theory. The second source of support for the convergence hypothesis rests on the belief that public policies have been successful in channelling resources and opportunities to the poor in general and to depressed regions and distressed cities in particular. Central to this view is the contention that regional policies, together with the effects of social security payments and of public expenditure in general have been effective in reducing interregional inequality through generating employment and providing infrastructural improvements and social amenities; and that regional planning has softened the effects of the ongoing structural transformation of the economy.

The divergence hypothesis is based on the expectation of increasing spatial inequalities. It is supported by several major theoretical perspectives. The first stems from the work of Myrdal (1957) and Hirschmann (1958) and is based on the idea of vicious circles of 'cumulative causation'. These writers emphasise the drain of skill, enterprise and capital from declining regions and the ability of more prosperous regions to extend their market into less developed areas. This polarisation of economic activity, it is argued, is sustained by a series of cumulative advantages. Superior economic performance in growth regions provides the basis for better services. Better services, in turn, add further dynamic impulse; and so on. While it is acknowledged that some countervailing 'spread effect' may transmit growth impulses to the depressed regions, the cumulative product of these processes is an intensification of spatial inequality. The idea of an increasing geographical polarisation of prosperity is also associated with Marxist and neo-Marxist theory. This perspective is based on the argument that the dynamics of capital accumulation inevitably lead to some form of core-periphery system of relations in production and exchange: a system which results in uneven development and the progressive intensification of relative spatial disparities (Carney, Hudson and Lewis 1980).

The *status quo* hypothesis would be validated by a situation in which there was no significant change in the pattern of spatial inequality. One reason for expecting such an outcome would be the existence of serious imperfections in the 'natural' forces of convergence stressed by neo-classical economic theory. In particular, these imperfections may include (i) the deterioration of social overhead capital (roads, schools, hospitals, etc.) in disadvantaged areas to the extent that they represent a handicap to economic recovery, and (ii) the failure of more prosperous areas to be 'pulled back' by the effects of agglomeration diseconomies as a result of the ability of large corporations to externalise their costs. A further line of argument in favour of the perpetuation of regional and inter-urban inequality is provided by Galbraith (1975), who suggests that, because small producers tend to be geographically dispersed and hence unable to wield much influence in the corridors of government power, economic growth remains concentrated in the large metropolitan business corporations which have the capacity to 'organise' the economy to their own advantage. Similarly, Holland (1976b) argues that large corporations have become immune to state policies, thereby undermining the effectiveness of regional policy in redressing spatial disparities.

Finally, it should be noted that there is one other outcome which is conceivable, but which is not normally considered feasible over the short or medium-term. This covers the possibility of a reversal of fortunes, with the best-off regions becoming the worst-off, and vice versa. Such an outcome may be named the 'inversion' hypothesis; it is not supported, however, by any existing theoretical perspectives.

Taking the infant mortality rate as a general-purpose indicator of relative regional social well-being, it emerges that the *status quo* has been maintained at the regional level since the 1930s. Figure 7.2 shows the relative distribution of infant mortality-rates among the regions of Western Europe in 1930 and 1980 (excluding Greece, France and Spain, for which data are not available). Quite clearly, regions with relatively high levels of infant mortality in 1930 also tended to have relatively high levels in 1980. This is confirmed by regression analysis, which also indicates that, in overall terms, there was no significant change in the pattern of relative regional inequality. There were, however, a considerable number of regions whose experience was distinctively different from that of the majority. These are the regions which fall outside the confidence limits of the regression equation on Figure 7.2. Using the average rates of infant mortality in 1930 and 1980 as reference points, it is possible to identify six different categories:

(1) those which began the period with relatively high rates of infant mortality and whose relative position has deteriorated further ('increasingly disadvantaged');
(2) those which began the period with relatively low rates of infant mortality, but which by the end of the period had relatively high rates ('reversal to disadvantaged');
(3) those which began the period with relatively low rates of infant mortality, but which by the end of the period had near-average rates ('convergent advantaged');
(4) those which began the period with relatively low rates of infant mortality and whose position has improved significantly ('increasingly advantaged');

(5) those which began the period with relatively high rates of infant mortality, but which by the end of the period had relatively low rates ('reversal to advantaged'); and

(6) those which began the period with relatively high rates of infant mortality, but which by the end of the period had near-average rates ('convergent disadvantaged').

Figure 7.2: Changes in Regional Rates of Infant Mortality, 1930-80

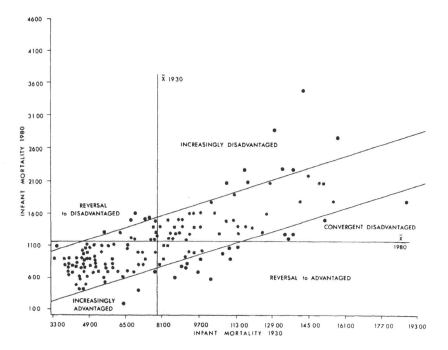

Altogether, these atypical cases represent over 20 per cent of the total, suggesting that, overall *status quo* notwithstanding, the dynamics of regional change in social well-being have produced significant tendencies towards convergence, divergence and – rather surprisingly – inversion.

Table 7.1, which lists the regions in each of the six residual categories, shows that there is a considerable degree of spatial coherence in their distribution. The eight regions in the increasingly disadvantaged category, for example, are all from the periphery of Mediterranean Europe, while the three increasingly advantaged regions are all from Scandinavia. When the exercise is repeated for the more recent period of 1950-80 (this time including the regions of France and Spain), similar results emerge. Once again, the overall relationship reflects the

Table 7.1: Regional Change in Infant Mortality, 1930-80:
Residual[a] Regions

1930-80	1950-80
(1) Increasingly disadvantaged	
Minho (Portugal)	Minho (Portugal)
Trás-os-Montes e Alto Douro (Portugal)	Trás-os-Montes e Alto Douro (Portugal)
Beira Alta (Portugal)	Beira Alta (Portugal)
Beira Littoral (Portugal)	Beira Littoral (Portugal)
Ribatejo (Portugal)	Ribatejo (Portugal)
Campania (Italy)	Asturias (Spain)
Calabria (Italy)	Vascongadas (Spain)
(2) Reversal to disadvantaged	
Vorarlberg (Austria)	Bremen (W. Germany)
Vienna (Austria)	Baleares (Spain)
West Berlin (W. Germany)	North (UK)
South West (UK)	East Midlands (UK)
Sydlige Jylland (Denmark)	West Midlands (UK)
	Sydlige Jylland (Denmark)
	Bornholm (Denmark)
(3) Convergent advantaged	
Oppland (Norway)	East Anglia (UK)
	South East (UK)
(4) Increasingly advantaged	
Sjaelland (Denmark)	Gotland (Sweden)
Gotland (Sweden)	Värmland (Sweden)
Häme (Finland)	Häme (Finland)
	Mikkeli (Finland)
(5) Reversal to advantaged	
Nordlige Jylland (Denmark)	Hainaut (Belgium)
Østlige Jylland (Denmark)	Liège (Belgium)
Lolland-Falster (Denmark)	Limbourg (Belgium)
Anvers (Belgium)	Namur (Belgium)
Limbourg (Belgium)	Kuopio (Finland)
Namur (Belgium)	Oulou (Finland)
Oost Vlaanderen (Belgium)	Lappi (Finland)
Kymi (Finland)	
Mikkeli (Finland)	
Kuopio (Finland)	
Oulu (Finland)	
(6) Convergent disadvantaged	
Burgenland (Austria)	Burgenland (Austria)
Estremadura (Portugal)	Oberösterreich (Austria)
Lombardia (Italy)	Lombardia (Italy)
Venezia Giulia (Italy)	Venezia Giulia (Italy)
Veneto (Italy)	Castilla La Nueva (Spain)

Note: a. Significant at 5 per cent.

status quo, though there are some regions whose relative position is significantly different. As Table 7.1 shows, the divergent regions are once again from the Mediterranean periphery and Scandinavia respectively. The composition of the other categories provides some intriguing hints as to the nature of spatial change. Disadvantaged but improving regions tend to be core regions of peripheral countries: Lombardy, New Castille, Estremadura; while those regions which have been transformed from a disadvantaged to a relatively favoured position have been from the economically intermediate regions of northwestern Europe: southern Belgium, northeastern Jutland and southern Finland. Meanwhile, two of the casualties of geopolitics, West Berlin and Vienna, together with some of the industrial regions of Britain, are among the representatives of inverted fortunes, once favoured but now disadvantaged. These atypical cases are interesting; yet they should not be pursued too far for their own sake. What is important to note here is that (i) while the absolute degree of inequality has decreased, the relative inequality of European regions remains much the same, and (ii) while the *status quo* has predominated, the nature of regional change has been complex, resulting in significant changes in the relative position of a variety of regions.

The Changing Social Welfare Environment

Whatever the causal processes, then, problems of inequality and relative regional disadvantage have persisted into the era of mature welfare states. Yet, as we have seen, the activities of central and local governments now influence, in one way or another, every aspect of social well-being. Moreover, the opportunities and life chances of the poor in particular have become increasingly dependent on patterns of public expenditure and provision, taxation and policy. At this point, therefore, it is useful to consider the role of the state in relation to the social welfare environment. This, of course, is an issue which has exercised political scientists a good deal, and which geographers have recently come to recognise as one of the principal determinants of the human landscape in Western societies (Clark and Dear 1981; Dear and Clark 1978; Johnston 1982; Kirby 1982a). In relation to the Western European scene, two major perspectives can be identified: 'liberal-democratic' and neo-Marxist. Briefly, the liberal-democratic tradition views the state as an autonomous set of politically neutral institutions. It is 'up for grabs', to be 'captured' by elected parties or coalitions and used as an instrument for their own objectives (Scase 1980). Within Western Europe there have emerged two rather different translations of this tradition. On the one hand there is the 'conservative-liberal' translation which stresses that the chief function of the state is 'facilitative'. That is, it provides the general framework within which business can operate, intervening under exceptional circumstances to ameliorate problems created by the market economy. On the other hand there is the 'social-democratic' translation which tends to regard the state as an institution which can be 'captured' and used not only for adjustments within a market economy but also for the long-term transformation of the market economy itself. Hence the concepts of 'welfare capitalism', 'democratic socialism', 'parliamentary socialism' and so on which have emerged from the social democratic parties of Western Europe.

The neo-Marxist tradition, in contrast, claims that the state cannot be 'captured' and used as a means of transforming society. Rather, it is argued that the state, whatever the motives of particular government regimes, inevitably serves as a key mechanism through which the needs of capital (and, hence, those of the dominant class/region) are served (Causer 1978). Thus, for example, the expansion of the state's roles in terms of the provision of health and social welfare facilities, the sponsorship of technological research and the creation of public enterprises are interpreted not in terms of social concern or economic management, but as evidence of the way in which the state has been used to 'underwrite' the costs – direct and indirect – of capitalist production. Much the same argument has been applied to the operation of regional and local government (see, for example, Kirby 1982b). Such interpretations recognise, however, that there will rarely be a 'fit' between the interests of capital and the activities of the state (or of the 'local state'). Apart from anything else, capital itself does not have a common interest: the needs of small, local businesses conflict with those of large, multinational enterprises, and so on. It is also recognised that the activities of the state will differ considerably in a 'dependent' or 'peripheral' economy compared with one in which the development of large-scale monopoly capitalism has reached an advanced stage. Indeed, Marxian frameworks of analysis provide for (and are in fact based on) a sophisticated degree of contradiction and conflict. The empirical evidence relating to state and local-state activity is fragmentary, however, and therefore open to interpretation. The following paragraphs present a brief review of the most significant aspects of the contemporary social welfare environment in relation to patterns of relative inequality.

National Trends

The first point to note in this context is that, in general, the bulk of the postwar expansion of social welfare programmes have not been aimed at low-income groups. Rather, it is 'average' employees and their families who have benefited most from improvements in income maintenance and social security. Expenditures have thus tended to redistribute income horizontally within similar income groups and, until recently at least, have done little to relieve poverty among the low-paid, the elderly or single-parent families (George and Lawson 1980; OECD 1976c). Even in West Germany, which has gone further than other countries in developing generous social welfare programmes, there is still a significant minority of elderly and disabled people in poverty, mainly widows of unskilled or semi-skilled workers and others with very low pensions because of their failure to fulfil the necessary insurance contributions (Lawson 1980). The European Community's programme of pilot schemes and studies on poverty (*Europe Against Poverty*) has highlighted three major gaps in European welfare states: (i) the educational neglect of lower working-class children, (ii) the inadequacy of pensions for women and those with an interrupted record of social insurance contributions, and (iii) the housing needs of immigrants and ethnic minorities (Room 1982). In addition, the programme has pointed to the ways in which the expansion of the welfare state has unintentionally created *new* forms of deprivation and disadvantage: the socially incapacitating effects of many institutionalised forms of health care, for example, and the socially disruptive effects of urban redevelopment schemes and housing projects.

Inasmuch as the victims of these shortcomings are unevenly distributed geographically, so the operation of the welfare state has involved a spatial differential. Moreover, this differential can be compounded by other factors. In West Germany, for example, social welfare programmes have been financed in part by the migrant-worker populations of industrial cities whose take-up of welfare rights has, for a variety of reasons, been low (Lawson 1980). The overall effect has thus been a net transfer of benefits to more prosperous citizens and localities within the system. From the liberal-democratic perspective, these shortcomings of the welfare state can be interpreted as part and parcel of the democratic process: as governments try to do more and more for a wider and wider spectrum of society, tensions and contradictions create 'residual' inequalities (Daudt 1977). From the neo-Marxist perspective, they are a natural component of welfare capitalism, since capitalism loses both its driving force and its *raison d'être* without the persistence of substantial inequalities (Gough 1982).

Yet, as the shortcomings of European welfare states began to emerge as serious theoretical and political issues, they were compounded and complicated by the change in the economic environment which occurred in the mid-1970s. The dilemma facing most governments was that the deepening economic recession accentuated the vulnerability of more and more people while making it increasingly difficult – economically and politically – to finance existing social welfare programmes. As a result, it has been suggested that Europe's welfare states have entered a new stage of reformulation and retrenchment (Heclo 1981; Navarro 1982), and a good deal of attention has been given to the 'new conservatism' associated with the orientation of central and local governments in the wake of recession. Electoral victories of right-of-centre parties in the Netherlands, Sweden and the United Kingdom (along with those in the United States, Australia and New Zealand) were associated with an ideological stance based on the belief that the welfare state had not only generated unreasonably high levels of taxation, budget deficits, disincentives to work and save, and a bloated class of unproductive workers, but also that it may have fostered 'soft' attitudes toward 'problem' groups in society. As Logue (1979) points out, the electoral appeal of this ideology can largely be attributed to the *success* of welfare states in banishing the spectre of material deprivation through illness, bereavement, unemployment and old age: consequently, the priority accorded to welfare programmes has receded (though the logic and, critically, the costs of maintaining them have not).

The restructuring of the welfare state has been most pronounced in the United Kingdom, where the Conservative Government has embarked on a programme of privatisation in housing, health and education, accompanied by cuts (some absolute, some relative) in higher education, in social welfare programmes for the unemployed, the disabled, the elderly and strikers' families, and in the regional policy budget. Meanwhile, closer controls on local authority expenditure by the central government have precipitated corresponding cuts at the local level, particularly in depressed towns and cities where the incidence of need for welfare services is high but local fiscal resources are low (Glennerster 1980; Knox and Kirby 1984). Other countries have responded in different ways to the post-1973 economic recession. Indeed, there appear to have been no systematic relationships between economic performance, political control, policy orientation and policy output. As Table 7.2 shows, Ireland, Italy, Spain and the United

Kingdom all turned in an unstable economic performance, with low growth-rates and high rates of inflation, declining competitiveness, high rates of unemployment and small increases in welfare expenditure. At the other end of the league table, Austria and Norway, along with Japan, enjoyed relatively high growth-rates, low rates of inflation, stable or increasing competitiveness, low unemployment and at least a moderate expansion in social security programmes. Yet in terms of policy orientation over the same period the pattern is quite different (OECD 1981a; Schmidt 1983). The United Kingdom, following the United States and Canada, gave priority to economic accumulation and the control of inflation at the expense of full employment and social welfare programmes. In contrast, Denmark, Luxembourg and Sweden gave priority to full employment and welfare provision at the expense of economic growth and competitiveness. Schmidt (1983) attributes this to the strength of welfare state bureaucracies and their clientele in these countries, but concedes that such strength does not always result in an expansive social welfare policy. In Finland, the Netherlands and West Germany, for example, the main thrust of central government policy during economic recession has been towards the modernisation of the economy, coupled with a modest growth in welfare expenditure through a corresponding expansion of public debt. In these circumstances, the net result has been a marked increase in unemployment, relieved by relatively generous but increasingly restrictive unemployment benefits. In other countries, other factors have had a decisive influence. In Switzerland, for example, the good economic performance and low rate of unemployment were the result of a big reduction in the labour force, mainly at the expense of foreign workers. Meanwhile, Portugal has been exceptional in achieving a marked increase in social welfare expenditure while experiencing low rates of economic growth, high rates of inflation and high rates of unemployment – all products of social revolution during economic recession (OECD 1981b). In short, it is overly simplistic to characterise the post-1973 era as one of reformulation or retrenchment under the influence of a 'new conservatism'. Rather, the variability of central government responses to economic recession seems to be the result of complex interactions between the previous level of national economic strength, the nature of the political culture, and the relative strength of extra-parliamentary forces such as trades unions and business organisations (Castles 1982; Schmidt 1983).

Table 7.2: Economic Performance and Policy Outcomes in 23 OECD Nations 1974-80

	Rank-order of economic performance 1979-80[a]	Labour-market performance and rates of unemployment 1973-80[b]	Growth of social security expenditure in percentage points of GDP 1973-9	Public debt as % of GDP 1973	1979
Australia	8	average (4.5)	2.6	—	
Austria	5	good (1.8)	2.7	17	31
Canada	6	average (7.2)	0.8	—	
Denmark	15	average (6.0)	4.7	16	36
Finland	12	average (4.6)	1.6	6	15
France	11	bad (4.8)	4.9	13	16
Germany (West)	4	bad (3.2)	2.7	18	29
Greece	14	good (1.9)	2.1	—	
Iceland	22	good (0.5)	−1.5	—	
Ireland	13	bad (7.0)	2.1	60	91
Italy	18	bad (6.7)	0.3	33	69
Japan	1	good (1.9)	5.8	13	36
Luxembourg	—	good (0.5)	7.5	28	25
Netherlands	9	average (4.1)	7.1	43	44
New Zealand	19	good (1.9)	—	—	
Norway	3	good (1.8)	1.6	33	53
Portugal	21	bad (7.0)	5.4	—	
Spain	20	bad (6.1)	2.5	—	
Sweden	16	good (1.9)	5.7	36	50
Switzerland	7	average (0.5)	2.5	25	30
United Kingdom	17	bad (5.4)	2.5	73	62
USA	2	average (6.7)	1.6	49	45

Notes: a. Rank 1 = 'best performance'; rank 22 = 'worst performance'.
b. Data in parentheses are rates of unemployment (average 1974 to 1980). 'Bad' means a rate of unemployment which is greater than 2% and a decline in the level of employment between 1973 and 1980. 'Good' means a low rate of unemployment (less than 2%) and an increase in the level of employment. 'Average' means either a low rate of unemployment and a decrease in the level of employment (for example, the Swiss case) or a high rate of unemployment and an increase in the level of employment (for example, USA).
Source: Schmidt (1983).

The Local Fiscal Crisis and Patterns of Public Provision

In most Western European countries the primary agency for actually delivering the benefits and services of the welfare state is local government, and it is this level of government which is therefore a principal determinant of 'territorial distributive justice' − both between and within local administrative units. The ability of some local governments to act effectively as the custodians of local social welfare, however, has

Table 7.3: Standardised Financial Accounts for Selected European Cities

City	Composition of Expenditure (%)											Tax receipts % derived from local taxes on:						Debt — Medium and Long-term Debt per capita[a]
	General administration	Interest	Security police, justice	Education	Public health	Social welfare	Housing and town planning	Public sanitation	Culture, leisure, sport	Traffic and transport	Economic services	Goods and services	Income and profits	Property	Mixed sources	Payroll	Capital	
Aalborg (Denmark)	4	1	0	15	0	65	3	1	3	3	3	0	34	2	0	0	0	702
Auxerre (France)	9	8	3	14	0	6	9	5	21	21	1	8	0	6	28	0	1	577
Aveiro (Portugal)	37	2	2	4	0	1	2	15	8	8	7	10	50	6	0	0	0	10
Cork (Ireland)	7	19	7	2	3	3	28	9	8	9	6	0	0	47	0	0	0	464
Coventry (UK)	1	17	0	51	0	8	14	4	4	2	4	0	17	13	17	0	0	766
Düsseldorf (W. Germany)	15	8	3	8	5	16	12	7	11	10	2	1	32	5	1	13	1	1077
Esbjerg (Denmark)	16	4	1	19	2	43	9	4	4	5	2	3	5	1	0	0	0	892
Florence (Italy)	2	33	3	16	3	4	1	12	3	3	7	1	50	60	0	0	12	901
Gothenburg (Sweden)	10	3	10	14	35	18	8	1	5	10	2	0	59	0	0	0	0	902
Lausanne (Switzerland)	8	9	7	18	3	6	3	10	12	13	1	0	0	8	0	0	3	1936
Limerick (Ireland)	18	21	4	2	1	3	28	9	6	11	4	6	0	46	0	0	0	474
Linz (Austria)	9	2	4	6	20	13	8	3	10	10	2	1	59	4	37	11	0	125
Lugano (Switzerland)	7	14	3	9	25	8	1	5	4	11	4	17	0	1	0	0	1	3460
Marseille (France)	12	8	5	13	0	9	5	9	7	36	2	1	14	4	26	0	1	572
Munich (W. Germany)	12	5	15	13	13	14	7	8	6	18	1	3	6	6	22	0	1	452
Namur (Belgium)		2	3	11	12	10	10	6	9	11	0	7	0	12	1	1	0	499
Orleans (France)		9	3	19	0	7	6	6	24	12	1	56	14	9	31	1	1	566
Palermo (Italy)	8	50	7	3	1	1	0	14	0	15	3	4	38	13	1	0	0	399
Porto (Portugal)	44	4	2	2	0	2	32	1	1	4	0	1	20	4	0	0	0	51
Siegen (W. Germany)	24	9	1	15	1	12	7	7	6	11	6	1	49	5	2	12	0	655
Stockholm (Sweden)	4	8	2	20	0	29	15	0	9	9	4	1	51	0	11	0	0	1444
Umeå (Sweden)	4	5	6	34	0	20	15	2	6	8	3	1		0	0	0	0	738
Utrecht (Netherlands)	5	11	10	12	3	21	22	4	9	7	1	1		4	0	0	0	626
Zürich (Switzerland)	9	10		18	9	13	3	5	7	12	2	0	0	53	0	0	0	3396

Note: a. In European Units of Account.
Source: Council of Europe (1981).

been severely limited by the critical state of their finances. Over the past 20 or 30 years the expenditures of local governments throughout Western Europe have been escalating even more rapidly than public expenditure as a whole, while local sources of income have expanded at a slower rate, thus making local governments more dependent on central government grants. In many cases, however, these have not been sufficient to meet increases in spending, so that local authorities have had to resort to borrowing. This, in turn, has increased their debt charges, leaving proportionally less money to pay for essential services and so initiating a vicious cycle of deficits, loans and bigger deficits (see Table 7.3).

It should be stressed that while this 'local fiscal crisis' is widespread, it is not to be found in every country; and, even where it does exist, it rarely reaches full crisis proportions. Nevertheless, the problem is serious and it is growing. As Newton *et al*. (1980) observe, the likes of Birmingham, Brussels, Copenhagen, Frankfurt, Liverpool, Milan, Munich, Naples, Palermo and Rome have all had to resort to fiscal measures more stringent than anything since war-time economies, and:

> Other cities join the list as each month passes. If it is not a cut in public services, the cancellation of new plans, or a reduction in the workforce, it is a large increase in local taxes or an equally unpopular increase in charges such as bus fares or rents for public housing. Nor is it just the big cities which are in trouble, for some towns, villages and rural areas are also facing difficulties. (p. 3)

In Rome, local officials talk of the 'monthly miracle' which enables the city to pay its operating expenses and to avoid bankruptcy yet again. In Liège, local government employees were put on subsistence wages when the city went bankrupt in 1983 and had to negotiate a massive loan from the central government in order to bail itself out.

Contrary to popular opinion, the financial problems of local authorities are not the result of inefficiency, corruption, extravagance, ineptitude and profligacy. Nor are they simply a reflection of national economic conditions. Problems are more severe in Germany, for example (Schäfer 1981), than in Sweden (Hansen 1981), where economic growth has been rather more modest. Rather, they are the outcome of an increasing disparity between the responsibilities and commitments of local governments and their discretionary powers (Sharpe 1980). A comparative study of local government finances in six European countries (Denmark, Italy, Norway, Sweden, the United Kingdom and West

Germany) suggests that the problem boils down to a 'local resource squeeze' arising from the conflict between, on the one hand, those forces contributing to rising expenditure and, on the other, those which limit the amount of local income (Newton *et al*. 1980). The study identified six main sets of factors which have contributed to rising local authority expenditures:

(1) *An expansion of existing services* Partly because social, economic and demographic changes have increased the size of those groups which typically make the biggest demands on local government services. The increasing number of retired persons, for example, requires a wider range of amenities and services, while the increasing number of working mothers has created a demand for day-care facilities. Changes in economic structure have required an expansion of educational programmes; and increasing private affluence has required increasing public expenditure: private cars need public roads and parking lots, for example.

(2) *Increases in the quality of services provided* This is seen as a function of rising expectations and the increasing disposition towards the idea of social justice, both of which are associated with increasing affluence. Often, this has been translated by central governments into sets of minimum standards for service provision: housing-construction codes, teacher-pupil ratios, and so on. The increasing influence of local government bureaucracies has also contributed to improvements in service quality as professional planners, housing managers, environmental health officers, social workers and so on press for improvements and innovations within their own particular domain.

(3) *Increases in the range of services* While there has been a general tendency towards centralisation, with local governments losing some services to central governments (Chapter 3), the range of local authority services has been extended considerably in the areas of consumer protection, environmental protection, inner-city regeneration, recreation, and in the provision of services for special groups such as migrant workers and the handicapped. This is seen by Newton *et al*. (1980) as a result of local governments' natural ability to be more responsive to local needs and demands.

(4) *The replacement of social capital* This is simply the legacy of the early start experienced by some cities and regions in terms of 'modern' economic development. As a result, they now find themselves with a certain amount of outworn and inadequate social capital in the form of old schools, bad housing and poor roads. In addition, some areas have

had to spend heavily on reconstruction in the wake of the Second World War.

(5) *The increased costs of services* The point here is that rates of inflation have had a particularly adverse effect on local government finances because local government services are both more capital-intensive and more labour-intensive than higher levels of government.

(6) *The costs of urbanisation* The increasing urbanisation and suburbanisation of most European countries is associated with several factors which induce higher levels of local government expenditure. Not only are the initial costs of urban development rather high – new sewers, water supplies, roads, schools – but as population density increases so do negative externalities such as traffic congestion, environmental degradation and deviant behaviour, all of which require local government services. In addition, as cities increase in size and density, so they generate needs for collective services such as street lighting, public toilets, public transport, and so on. They also begin to produce, attract and retain a disproportionate number of service-dependent households and to suffer from the effects of agglomeration diseconomies. Larger metropolitan areas also act as central places for very large hinterlands, thus providing a variety of amenities and services – from arts centres to zoos – which are used by visitors and commuters as well as by their own population.

In relation to the constraints on local government incomes, Newton *et al*. identify four main factors:

(1) *The lack of buoyancy of some local taxes* In essence, the problem here is that central governments tend to take the greater share of those taxes which best match the general rate of inflation (such as income tax, value-added tax, and payroll tax), leaving local governments to rely on land and property taxes (Table 7.3), which are less sensitive to inflation and more difficult to manipulate or adjust.

(2) *The visibility of some local taxes* The nature of local taxes often means that they are paid in a lump sum, and it is suggested that this makes local taxes politically difficult to increase.

(3) *The political sensitivity of local financial matters* It is argued that the prominence given to budgetary exercises in local politics tends to discourage increases in taxation, and that the frequency of local government elections tends to have the same effect, since local councils can rarely increase local taxes without facing the electorate soon afterwards.

(4) *Imbalances between needs and resources* There is often an inverse relationship between local needs and the magnitude of the local tax base. Indeed, the relationship has tended to intensify as a result of certain aspects of Europe's changing geography. The localisation of industrial, commercial and administrative activity, for example, has increased the tax base of more prosperous cities and regions, while metropolitan decentralisation has resulted in a net transfer of affluent and relatively undemanding taxpayers to small towns and suburban jurisdictions and a proportional increase in the less affluent and more service-dependent populations of central city areas.

Newton *et al.* found that the intensity of the local resource squeeze varies a good deal from country to country. Thus:

> Denmark and Sweden are in good financial condition, with debts and debt charges falling, expenditure rising, and no signs of serious financial shortages. Norway is not quite so fortunate but, nevertheless, has few major problems. A little worse is Germany where debts, debt charges and deficits have been growing and where expenditure, particularly capital expenditure, has had to be cut. Still worse is the UK where spending has been cut in real terms in some authorities because of a fairly severe financial shortage which has been accompanied by rising debt charges and increasing deficits. Worst of all is Italy where some communes are in a real state of crisis. Debts and debt charges are huge, and accumulating at an alarming rate. (1980: 24)

Variations *within* each country are also considerable, with the most striking and consistent differential being the one between larger cities and the rest. Even the large cities of the Scandinavian countries face more severe financial problems than their respective central governments, emphasising the suggestion that the 'urban resources squeeze' is more acute than the 'local resources squeeze'.

Quite clearly, these trends are of major significance in terms of geographical variations in the social welfare environment. The actual geography of local authority 'performance' in providing services and amenities remains somewhat unknown, however. The variability of local expenditure profiles is illustrated by Table 7.3. Spending on social welfare, for example, is much more prominent in the accounts of Danish, Dutch and Swedish cities than in Irish, Portuguese and Swiss cities. Meanwhile, Marseille is notable for the exceptionally high proportion of its total budget allocated to traffic and transport, as are Auxerre and Orleans for their expenditure on culture, leisure and sports facilities. Interest repayments dominate the budgets of Florence and Palermo; while the administrative costs of local government loom large in the budgets of Aveiro, Düsseldorf, Florence, Linz, Munich, Namur, Porto and Siegen. The sample of standardised accounts from which these data are drawn is too small, however, to test for systematic variations by region, urban size or economic base. Nor is it possible to gauge local authority 'performance' by matching these expenditure profiles to measures of need. We do know from the results of a number of individual studies, however, that local authority performance falls considerably short of achieving territorial distributive justice. Take, for example, the geography of local authority provision in the United Kingdom. There is now a considerable literature detailing the disparities which exist between need and provision across a wide spectrum of welfare programmes and at a variety of spatial scales (see, for example, Coates and Rawstron 1971; Davies, Barton and McMillan 1972; Kirby 1979; Pinch 1979; Pyle 1976). The two extremes of local authority performance are characterised by, on the one hand, areas where a declining industrial base is associated with poor housing and environmental conditions, a youthful, working-class population, and Labour Party control. Within their total budget they spend lower than average amounts on roads, police and parks, but higher than average amounts on health, housing, mothers and young children, and primary special education. Their expenditure on fire services and secondary education is average or above, while that on planning tends to be below the average. In general, therefore, they spend more on socially ameliorative and redistributive services but less on 'caretaker' services. At the other end of the scale are the relatively affluent residential/commercial communities with predominantly ageing, middle-class populations which consistently return Conservative-controlled councils. Their pattern of spending is reversed, in that they spend more than average on public goods such as roads, and less than average on divisible goods such as housing, medical services, personal social services and education (Newton 1976). Between these two extreme stereotypes, however, are various patterns of expenditure which are the product of complex interactions between (i) traditional local priorities (the emphasis on public housing in Scottish cities, for example, and the high priority given to education in Welsh authorities), (ii) local fiscal potential, (iii) the political balance of the council, (iv) the intensity of local needs for different services, and (v) the size of the authority and its degree of centrality in the national urban system (Alt 1971; Bennett 1982a; Boaden 1971; Danziger 1978; Newton 1984). In short, both the size and the composition of the 'social wage' depend very much on location, but there are no simple relationships with which to predict the outcome. Similar conclusions have been drawn from parallel research in other parts of Europe (Aiken and Depré 1980; Fried 1971, 1974, 1976; Ginsburgh and Pestieau 1981; Jackman 1980; Preteceille 1981). A good deal of responsibility for territorial distributive justice therefore falls on central governments. Yet there is little evidence that they are able – or willing – to effectively compensate for the idiosyncrasies of local government spending. In the United Kingdom, for example, the central government has a responsibility to assist local authorities with particularly high spending needs and/or low resource bases through the Rate Support Grant (RSG). For much of the postwar period, however, the operation of the RSG tended to favour the shire counties at the expense of metropolitan authorities and the inner London boroughs. For brief periods (with a Labour government) – 1974-5 and 1980-81 – metropolitan authorities were treated better than the shires by the RSG. More recently, however, the RSG has been used by the government as a device with which to sanction 'profligate' local authorities, with the result that the advantage has shifted back to the shires (Bennett 1981, 1982b; Johnston 1979b). Viewed together with the geographical implications of central government responses to recession, the apparent ineffectiveness of many regional economic policies, and the differential impact of the local resource squeeze, the prognosis must be that the social geography of Western Europe is likely to continue to be characterised by substantial disparities not only in affluence and opportunity, but also in the various components of the social wage which are increasingly important to the overall social well-being of communities and regions.

REFERENCES

Aarebrot, F.H. (1982) 'On the Structural Basis of Regional Mobilisation in Europe', in B. De Marchi and A.M. Boileau (eds), *Boundaries and Minorities in Western Europe*, Franco Angeli, Milan, pp. 33-92

Aase, A. and B. Dale (1978) *Levekår i Storby*, Norges Offentlige Utredninger 1978: 52, Universitetsforlaget, Oslo

Abrams, M. (1979) 'The Future of the Elderly', *Futures*, *11*, 178-84

Aceves, J.B. (1976) 'Forgotten in Madrid: Notes on Rural Development Planning in Spain', in J.B. Aceves and W.A. Douglas (eds), *The Changing Faces of Rural Spain*, Schenkman, New York, pp. 173-96

Agnew, J.A. (1978) 'Political Regionalism and Scottish Nationalism in Gaelic Scotland', *Canadian Review of Studies in Nationalism*, *6*

Aiken, M. and Depré, R. (1980) 'Policy and Politics in Belgian Cities', *Policy and Politics*, *8*, 73-106

Albers, G. (1980) 'Town Planning in Germany: Change and Continuity under Conditions of Political Turbulence', in G.E. Cherry (ed.), *Shaping an Urban World*, Mansell, London, pp. 145-60

Aldcroft, D.H. (1978) *The European Economy 1914-1970*, Croom Helm, London

Aldskogius, H. (1978) 'Leisure Time Homes in Sweden', *Current Sweden*, *195*, 1-27

Allardt, E. (1981) 'Experiences from the Comparative Scandinavian Welfare Study, with a Bibliography of the Project', *European Journal of Political Research*, *9*, 101-12

Allen, K. (1981) 'Some Recent Trends in European Regional Incentive Policy', *Built Environment*, *7*, 211-221

Almeida, L.T. (1980) 'The Use of Socio-economic Indicators to Group European Countries', in J. de Bandt, P. Mándi and D. Seers (eds), *European Studies in Development*, Macmillan, London

Alt, J.E. (1971) 'Some Social and Political Correlates of County Borough Expenditures', *British Journal of Political Science*, *1*, 48-62

Ambrose, P. (1974) *The Quiet Revolution*, Chatto & Windus, London

Amman, A. (1981) 'The Status and Prospects of the Aging in Western Europe', *Eurosocial Occasional Papers*, *8*, Vienna

Andereisen, G. (1981) 'Tanks in the Streets: The Growing Conflict over Housing in Amsterdam', *International Journal of Urban and Regional Research*, *5*, 83-95

Andersen, O. (1979) 'Denmark' in W.R. Lee (ed.), *European Demography and Economic Growth*, Croom Helm, London, pp. 79-122

Andorka, R. (1978) *Determinants of Fertility in Advanced Societies*, Methuen, London

Appleyard, D. (1979) *The Conservation of European Cities*, MIT Press, Cambridge, Mass.

Après-Demain (1980) 'La France Vieillit' (Special issue), *220*, 3-40

Arcangeli, F. (1982) 'Regional and Subregional Planning in Italy', in R. Hudson and J.R. Lewis (eds), *Regional Planning in Europe*, Pion, London, pp. 57-84

——, Borzaga, C. and Goglio, S. (1980) 'Patterns of Peripheral Development in Italian Regions', *Papers and Proceedings of the Regional Science Association*, *44*, 19-34

Arensberg, C.M. and Kimball, S.T. (1940) *Family and Community in Ireland*, Smith, Dublin

Armstrong, W.H. (1978) 'Community Regional Policy: A Survey and Critique', *Regional Studies, 12*, 511-28

Arnold, T., Danieli, L. and Zacchia, C. (1981) 'Regional Concentration of Population and Inter-regional Migration in Europe, 1950-1970', in A. Kuklinski (ed.), *Polarized Development and Regional Policies*, Mouton, The Hague, pp. 411-506

Ashcroft, B. (1980) *The Evaluation of Regional Policy in Europe: A Survey and Critique*, Studies in Public Policy No. 68, Centre for the Study of Public Policy, University of Strathclyde, Glasgow

Asheim, B.T. (1978) *Regionale ulikheter; levekår*, Norges Offentlige Utredninger No. 3, Universitetsforlaget, Oslo

Ashford, D.E. (1978) 'The Limits of Consensus; The Reorganisation of British Local Government and the French Contrast', in S. Tarrow, P.J. Katzenstein and L. Graziano (eds), *Territorial Politics in Industrial Nations*, Praeger, NY, pp. 245-89

—— (1979) 'Territorial Politics and Equality: Decentralisation and the Modern State', *Political Studies, 27*, 71-83

Association of County Councils (1979) *Rural Deprivation*, Association of County Councils, London

Atkinson, A. (1975) *The Economics of Inequality*, Clarendon Press, Oxford

Aydalot, P. (1978) 'Regional Planning in France: A Tentative Evaluation', *L'Espace Géographique, 7*, 245-53

Ayton, J.B. (1976) 'Rural Settlement Policy: Problems and Conflicts', in P.J. Drudy (ed.), *Regional and Rural Development: Essays in Theory and Practice*, Alpha Academic, London

Backer, J.E. (1966) 'Norwegian Migration 1856-1960', *International Migration, 4*, 172-85

Bagnasco, A. (1982) 'Economia e società della piccola impresa', in S. Goglio (ed.), *Italia: centri e periferie*, S. Angeli, Milan

Ball, R. (1980) *Spatial Problems in an Expanded European Community*, Working Paper No. 78, Centre for Urban and Regional Studies, University of Birmingham

Banister, D.J. (1980) *Transport Mobility and Deprivation in Inter-Urban Areas*, Saxon House, Farnborough

—— (1983) 'Transport and Accessibility', in M. Pacione (ed.), *Progress in Rural Geography*, Croom Helm, London, pp. 130-48

Beale, C. (1977) 'The Recent Shift in United States Population to Nonmetropolitan Areas, 1970-1975', *International Regional Science Review, 2*, 113-22

Beijer, G. (1969) 'Modern Patterns of International Migratory Movements', in J.A. Jackson (ed.), *Migration*, Cambridge University Press, pp. 100-24

Bemmel, A.B. van (1981) 'A Welfare Geographical Approach of Living in Rural Areas: Some Remarks', Paper XI, European Congress for Rural Sociology, Helsinki

Benevolo, L. (1980) *The History of the City*, transl. G. Culverwell, MIT Press, Cambridge, Mass.

Bennett, R.J. (1981) 'The RSG in England and Wales 1967/8 to 1980/1: A Review of Changing Emphases and Objectives', *Geography and the Urban Environment, 4*, 139-91

—— (1982a) 'The Financial Health of Local Authorities in England and Wales: Resource and Expenditure Position 1974/5 to 1980/1', *Environment and Planning, 14A*, 997-1022

—— (1982b) 'The Financial Health of Local Authorities in England and Wales:

the Role of the RSG', *Environment and Planning*, *14A*, 1283-306

Bentham, G. and Moseley, J.J. (1980) 'Socio-economic Change and Disparities within the Paris Agglomeration: Does Paris Have an 'Inner-city' Problem?', *Regional Studies*, *14*, 55-70

Berger, M., Fruit, J.P., Plet, F. and Robie, C. (1980) 'Reurbanisation and the Analysis of Peri-urban Rural Space', *L'Espace Géographique*, *9*, 303-14

Berentsen, W.H. (1978) 'Austrian Regional Development Policy: The Impact of Policy on the Achievement of Planning Goals', *Economic Geography*, *54*, 115-34

Berg, L. van den, Drewett, R., Klaasen, L.H., Rossi, A. and Vijverberg, C.H.T. (1982) *Urban Europe: A Study of Growth and Decline*, Pergamon, Oxford

Bergman, S. (1979) 'The Future of Human Welfare for the Aged', in H. Orimo, K. Shimada, M. Iriki and D. Maeda (eds), *Recent Advances in Gerontology*, Excerpta Medica, Amsterdam, pp. 44-8

Bergman, T. (1970) 'Land Consolidation in Germany', in A.H. Bunting (ed.), *Change in Agriculture*, Duckworth, London, pp. 481-5

Berkner, L. and Mendels, F. (1978) 'Inheritance Systems, Family Structure and Demographic Patterns in Western Europe 1700-1900', in C. Tilly (ed.), *Historical Studies of Changing Fertility*, Princeton University Press, Princeton, pp. 209-24

Bernard, H.R. and Ashton-Vouyoucalos, S. (1976) 'Return Migration to Greece', *Journal of the Steward Anthropological Society*, *8*, 31-52

Berry, B.J.L. (1973) *The Human Consequences of Urbanisation*, Macmillan, London

Beyer, G. (ed.) (1961) *Characteristics of Overseas Migrants*, Government Printing Office, The Hague

—— (1980) 'The Second Generation of Migrants in Europe: Social and Demographic Aspects', *European Demographic Information Bulletin*, *11*, 49-72

Bianchi, G. (1979) 'Regional Planning in Italy: A Critical Review', *Planning Outlook*, *22*, 2-5

Bilmen, M.S. (1976) 'Educational Problems Encountered by the Children of Turkish Migrant Workers', in N. Abadan-Unat (ed.), *Turkish Workers in Europe 1960-1975*, Brill, Leiden, pp. 235-52

Biraben, J-N. and Duhourcau, F. (1973) 'Etudes: la redistribution géographique de la population de l'Europe occidentale de 1961 à 1971', *Population*, *28*, 1158-69

Birch, A.H. (1978) 'Minority Nationalist Movements and Theories of Political Integration', *World Politics*, *30*, 325-44

Bird, R.M. (1971) 'Wagner's "Law" of Expanding State Activity', *Public Finance*, *26*, 1-26

Bird, S.E. (1982) 'The Impact of Private Estate Ownership on Social Development in a Scottish Rural Community', *Sociologia Ruralis*, *22*, 43-56

Birnbaum, P. (1980) 'The State in Contemporary France', in R. Scase (ed.), *The State in Western Europe*, Croom Helm, London, pp. 94-114

Blackaby, F. (ed.) (1979) *De-Industrialisation*, Heinemann, London

Blackborn, A. (1972) 'The Location of Foreign-owned Manufacturing Plants in the Republic of Ireland', *Tijdschrift Voor Economische en Sociale Geografie*, *63*, 438-43

—— (1982) 'The Impact of Multinational Corporations on the Spatial Organisation of Developed Nations: a Review', in M. Taylor and N. Thrift (eds), *The Geography of Multinationals*, Croom Helm, London, pp. 147-57

Blacksell, M. (1981) *Postwar Europe: A Political Geography*, Hutchinson, London

Blackwood, L.G. and Carpenter, E.H. (1978) 'The Importance of Anti-urbanism in Determining Residential Preferences and Migration Patterns', *Rural Sociology*, *43*, 31-47

Blijstra, R. (1981) *Town Planning in the Netherlands Since 1900*, P.N. Van Kampen & Zoon NV, Amsterdam

Boaden, N. (1971) *Urban Policy Making*, Cambridge University Press, Cambridge

Boal, F. (1976) 'Ethnic Residential Segregation', in D. Herbert and R.J. Johnston (eds), *Social Areas in Cities*, vol. 1, Wiley, Chichester, pp. 41-79

— and Douglas, J.N.H. (eds) (1982) *Integration and Division: Geographical Perspectives on the Northern Ireland Problem*, Academic Press, London

Bobek, H. and Lichtenberger, E. (1966) *Wien: bouliche Gestalt und Entwicklung seit der Mitte des 19. Jahrhunderts*, Verlag Hermann Böhlaus, Graz-Köln

Böhning, W.R. (1972) *The Migration of Workers in the United Kingdom and the European Community*, Oxford University Press, London

— (1975) 'Some Thoughts on Emigration from the Mediterranean Basin', *International Labour Review*, *3*, 251-77

Bonneau, M. (1979) 'Récréation et tourisme en France', in C. Christians and J. Claude (eds), *Recherches de géographie rurale*, Bulletin de la Société Géographique de Liège, Liège, 779-90

Borzaga, C. and Goglio, S. (1981) 'Economic Development and Regional Imbalances: The Case of Italy, 1945-1976', *Dunelm Translations*, *6*, Durham

Boudeville, J.R. (1978) 'Les régions de villes et l'Europe', in J.H.P. Paelinck (ed.), *La Structure urbaine en Europe occidentale: faits, théories, modèles*, Saxon House, Farnborough, pp. 51-92

Bourgeois-Pichat, J. (1981) 'Demographic Change in Western Europe', *Population and Development Review*, *7*, 19-42

Bourne, L. (1975) *Urban Systems: Strategies for Regulation*, Oxford University Press, London

Boustedt, O. (1975) *Grundriss der Empirischen Regionalforschung Teil III Siedlungsstrukturen*, Schoedel, Hannover

Bowler, I.R. (1976a) 'The CAP and the Space-economy of Agriculture in the EEC', in R. Lee and P. Ogden, (eds), *Economy and Society in the EEC*, Saxon House, Farnborough, pp. 235-55

— (1976b) 'Recent Developments in the Agricultural Policy of the European Economic Community', *Geography*, *61*, 28-30

— (1979) *Government and Agriculture: A Spatial Perspective*, Longman, London

— (1983) 'Structural Change in Agriculture', in M. Pacione (ed.), *Progress in Rural Geography*, Croom Helm, London, pp. 46-73

Boyer, J.-C. (1980) 'Résidences Secondaires et "rurbanisation" en Région Parisienne', *Tijdschrift Voor Economische en Sociale Geografie*, *71*, 78-87

Braun, R. (1978) 'Early Industrialisation and Demographic Change in the Canton of Zürich', in C. Tilly (ed.), *Historical Studies of Changing Fertility*, Princeton University Press, Princeton, pp. 289-334

Breathnach, P. (1982) 'The Demise of Growth-centre Policy: the Case of the Republic of Ireland', in R. Hudson and J. Lewis (eds), *Regional Planning in Europe*, Pion, London, pp. 35-56

Bresso, M. (1980) 'Les paradoxes de la Politique Agricole Commune', *L'Espace géographique*, *9*, 173-82

Brettell, C.B. (1979) 'Emigrar para voltar: a Portuguese Ideology of Return Migration', *Papers in Anthropology*, *20*, 1-20

Brown, A.J. and Burrows, E.M. (1977) *Regional Economic Problems*, Allen & Unwin, London

Bruun, F. and Skovsgaard, C.-J. (1980) 'Local Self-determination and Central Control in Denmark', *International Political Science Review*, *1*, 227-44

Bryant, C.R. (1974) 'An Approach to the Problem of Urbanisation and Structural Changes in Agriculture', *Geografiska Annaler*, *56B*, 1-27

Bryden, J. and Houston, G. (1976) *Agrarian Change in the Scottish Highlands: the Role of the HIDB in the Agricultural Economy of the Crofting Counties*, Martin Robertson, London

Buck, T.W. and Atkins, M.H. (1976) 'The Impact of British Regional Policies on Employment Growth', *Oxford Economy Papers*, *28*, 118-32

Buckwell, A.E., Harvey, D., Thomson, K. and Parton, K. (1982) *The Costs of the Common Agricultural Policy*, Croom Helm, London

Buchanan, R.H. (1979) 'Colonisation and Landscape', in C. Christians and J. Claude (eds), *Recherches de géographie rurale*, Bulletin de la Société géographique de Liège, Liège, pp. 277-300

Bunce, M. (1982) *Rural Settlement in an Urban World*, Croom Helm, London

Burke, G. (1976) *Townscapes*, Penguin, Harmondsworth

Burtenshaw, D., Bateman, M. and Ashworth, G.J. (1981) *The City in West Europe*, Wiley, Chichester

Butler, D. and Stokes, D. (1969) *Political Change in Britain*, Macmillan, London

Buttel, F.H. (1980) 'Agricultural Structure and Rural Ecology: Toward a Political Economy of Rural Development', *Sociologia Ruralis*, *20*, 44-62

Buursink, J. (1977) *De Hierachie von Winkelcentra*, Institute de Geographie, Groningen

Calot, G. and Blayo, C. (1982) 'Recent Course of Fertility in Western Europe', *Population Studies*, *36*, 349-72

—— and Hecht, J. (1978) 'The Control of Fertility Trends', in Council of Europe, *Population Decline in Europe*, Arnold, London, pp. 178-96

Calvaruso, C. (1982) 'I Laboratori Migranti della CEE e i problemi dell'occupazione negli anni 1980', *Studi Emigrazione*, *19*, 363-85

Camagni, R. and Cappelin, R. (1981) 'European Regional Growth and Policy Issues for the 1980s', *Built Environment*, *7*, 162-71

Capo, E. and Fonti, G.M. (1965) 'L'exode rural vers les grandes villes', *Sociologia Ruralis*, *5*, 267-87

Caporaso, J.A. (1974) *The Structure and Function of European Integration*, Goodyear, Pacific Palisades, California

Carney, J.G. (1976) 'Capital Accumulation and Uneven Development in Europe: Notes on Migrant Labour', *Antipode*, *8*, 30-8

——, Hudson, R. and Lewis, J. (eds) (1980) *Regions in Crisis*, Croom Helm, London

Carson, R. (1969) *Silent Spring*, Penguin, Harmondsworth

Carstairs, A.M. (1980) *A Short History of Electoral Systems in Western Europe*, Allen & Unwin, London

Carter, E. (1979) 'Politics and Architecture: An Observer Looks Back at the 30s', *Architectural Review*, November, 324-5

Carter, H. (1983) *An Introduction to Urban Historical Geography*, Arnold, London

Caselli, G. and Egidi, V. (1981) *New Trends in European Mortality*, Council of Europe, Population Studies No. 5, Strasbourg

Castells, M. (1982) *Citizens: A Cross-Cultural Theory of Urban Social Movements*, Arnold/University of California Press, London

——, Cherie, F., Godard, F. and Mehl, D. (1974) *Sociologie des mouvements sociaux urbains*, Ecole de Hautes Etudes en Sciences Sociales, Paris

Castles, F.G. (1982) 'Politics, Public Expenditure and Welfare', in F.G. Castles (ed.), *The Impact of Political Parties*, Sage, Beverly Hills

Castles, S. and Kosack, G. (1973) *Immigrant Workers and Class Structure in Western Europe*, Oxford University Press, London

Causer, G. (1978) 'Private Capital and the State in Western Europe', in S. Giner and M.S. Archer (eds), *Contemporary Europe: Social Structures and Cultural Patterns*, Routledge & Kegan Paul, London, pp. 28-54

Cawley, M.E. (1979) 'Rural Industrialisation and Social Change in Western Ireland', *Sociologia Ruralis*, *19*, 43-59

Ceccarelli, P. (1982) 'Politics, Parties and Urban Movements: Western Europe', in N. Fainstein and S. Fainstein (eds), *Urban Policy Under Capitalism*, Sage, Beverly Hills, pp. 261-76

Centre for Advanced Land Use Studies (1981) *The Inner City and Local Government in Western Europe*, Research Unit, College of Estate Management, Centre for Advanced Land Use Studies

Chadelet, J-F. (1975) *Who Will Pay Our Pensions in 1990?*, Council of Europe Working Paper, Strasbourg

Champion, A.G. (1981) 'Population Trends in Rural Britain', *Population Trends*, *26*, 20-3

—— (1982) 'Rural-urban Contrasts in Britain's Population Change 1961-1981', in A. Findlay (ed.), *Recent National Population Changes*, Institute of British Geographers, Population Study Group, London

Chapman, G. (1976) *Development and underdevelopment in Southern Italy*, University of Reading, Department of Geography, Geographical Papers No. 41

Chesnais, J.C. and Sauvy, A. (1973) 'Progrès économique et accroissement de la population', *Population*, *28*, 843-57

Chevalier, M. (1981) 'Les Phénomènes néo-Ruraux', *L'Espace géographique*, *10*, 33-47

Chiotis, G. (1972) 'Regional Development Policy in Greece', *Tijdschrift voor Economische en Sociale Geografie*, *63*, 94-104

Christians, C. and Claude, J. (eds) (1979) *Recherches de géographie rurale*, Bulletin de la société géographique de Liège, Liège

Clark, J.R. (1975) 'Residential Patterns and Social Integration of Turks in Cologne', in R. Ekrane (ed.), *Manpower Mobility Across Cultural Boundaries*, F.J. Brill, Leiden

Clark, G. and Dear, M. (1981) 'The State in Capitalism and the Capitalist State', in M. Dear and A.J. Scott (eds), *Urbanization and Planning in Capitalist Society*, Methuen, London, pp. 45-61

Clason, C. (1980) 'The One-parent Family: the Dutch Situation', *Journal of Comparative Family Studies*, *11*, 3-16

Claval, P. (1980) 'Centre/Periphery and Space: Models of Political Geography', in J. Gottmann (ed.), *Centre and Periphery*, Sage, London, 63-72

—— (1984) 'Cities of Western Europe', in J. Agnew, J. Mercer and D. Sopher (eds), *The City in Cultural Context*, Allen & Unwin, London

Cloke, P. (1979) *Key Settlements in Rural Areas*, Methuen, London

—— (1980) 'New Emphases for Applied Rural Geography', *Progress in Human Geography*, *4*, 181-217

Clout, H.D. (1969), 'Second Homes in France', *Journal of the Town Planning Institute*, *55*, 440-3

—— (1972) *Rural Geography: An Introductory Survey*, Pergamon Press, Oxford

—— (ed.) (1975) *Regional Development in Western Europe*, Wiley, London

—— (1976) 'Rural-urban Migration in Western Europe', in H.D. Clout and J. Salt (eds), *Migration in Post-War Europe*, Oxford University Press, London, pp. 30-51

Coale, A. (1969) 'The Decline of Fertility in Europe from the French Revolution to World War II', in S. Behrman *et al.* (eds), *Fertility and Family Planning: A World View*, University of Michigan Press, Ann Arbor, pp. 3-24

Coates, B.E. and Rawstron, E. (1971) *Regional Variations in Britain*, Batsford, London

——, Johnston, R.J. and Knox, P.L. (1977) *Geography and Inequality*, Oxford University Press, London

Coleman, D.A. (1980) 'Recent Trends in Marriage and Divorce in Britain and Europe', in R.W. Hiorns (ed.), *Demographic Patterns in Developed Societies*, Taylor & Francis, London, pp. 83-124

Collins, D. (1983) *The Operation of the European Social Fund*, Croom Helm, London

Commins, P. (1980) 'Imbalances in Agricultural Modernisation – with Illustrations from Ireland', *Sociologia Ruralis*, *20*, 63-81

Commission of the European Communities (1975) *Men and Women of Europe*, Commission of the European Communities, Brussels

—— (1977) 'Community Regional Policy: New Guidelines', *Bulletin of the European Communities*, 2/77

—— (1979) *Europe Against Poverty: Second Report of the European Programme of Pilot Schemes and Studies to Combat Poverty*, Commission of the European Communities, Brussels

—— (1980a) *Reflections on the CAP*, COM(SO) 800 final, Brussels

—— (1980b) *The Community and Its Regions*, European Documentation 1/80, Office for Official Publications of the European Communities, Luxembourg

—— (1981) *The Regions of Europe: First Periodic Report on the Social and Economic Situation of the Regions of the Community*, Commission of the European Communities, Luxembourg

Condorelli, L. (1979) 'Italy' in Policy Studies Institute, *European Integration, Regional Devolution and National Parliaments*, European Centre for Political Studies, Policy Studies Institute, London, pp. 22-32

Connell, J. (1974) 'The Metropolitan Village: Spatial and Social Processes in Discontinuous Suburbs', in J.H. Johnson (ed.), *The Geography of Suburban Growth*, Wiley, London, pp. 77-100

Connor, W. (1973) 'The Politics of Ethnonationalism', *Journal of International Affairs*, *27*

Coppock, J.T. (ed.) (1977) *Second Homes: Curse or Blessing?*, Pergamon, Oxford

Costa, F.J. (1977) 'The Evolution of Planning Styles and Planned Change: the Example of Rome', *Journal of Urban History*, *3*, 263-94

Coulter, J. (1978) *Grid Square Census Data as a Source for the Study of Deprivation in British Conurbations*, Working Paper no. 13, Census Research Unit, Department of Geography, University of Durham

Council of Europe (1966) *Report on the Evolution of Local and Regional Structures*, Council of Europe, DOC 2110, 26/9, 1966

—— (1975) *The Development of Central, Regional and Local Finance Since 1950*, Study Series Local and Regional Authorities in Europe, Study No. 13, vol. 1, Council of Europe, Strasbourg

—— (1978) *Population Decline in Europe*, Arnold, London

—— (1980) *Reviving Rural Europe*, European Regional Planning Study Series 29, Strasbourg

—— (1981) *Standardised European Local Accounts: Comparative Analysis of the Accounts of European Towns*, Study No 21, Council of Europe, Strasbourg

—— (1982) *Recent Demographic Developments in the Member States of the Council of Europe*, Steering Committee on Population Studies, Council of Europe, Strasbourg

Cox, A. (ed.) (1982) *Politics, Policy and the European Recession*, Macmillan, London

Cresswell, R. (ed.) (1978) *Rural Transport and Country Planning*, Leonard Hill, London

Cribier, F. (1974) 'Retirement Migration in France', in L.A. Kosinski and R.M.

Prothero (eds), *People on the Move*, Methuen, London, pp. 361-74
Cuddy, M. (1981) 'European Agricultural Policy: The Regional Dimension', *Built Environment*, 7, 200-10
Curl, J.S. (1970) *European Cities and Society*, Leonard Hill, London
Daalder, H. and Mair, P. (eds) (1983) *Western European Party Systems*, Sage, London
Dahrendorf, R. (1972) 'A New Goal for Europe', in M. Hodges (ed.), *European Integration*, Penguin, Harmondsworth
—— (1982) *European Economies in Crisis*, Weidenfeld and Nicolson, London
Dahya, B. (1974) 'The Nature of Pakistani Ethnicity in Industrial Cities in Britain', in A. Cohen (ed.), *Urban Ethnicity*, Tavistock, London, pp. 77-118
Damette, F. (1980) 'The Regional Framework of Monopoly Exploitation: New Problems and Trends', in J. Carney, R. Hudson and J. Lewis (eds), *Regions in Crisis*, Croom Helm, London, pp. 76-92
Daniels, P. (1982) *Service Industries: Growth and Location*, Cambridge University Press, Cambridge
Danziger, J.N. (1978) *Making Budgets*, Sage, Beverly Hills
Da Silva, M. (1975) 'Modernisation and Ethnic Conflict: the Case of the Basques', *Comparative Politics*, 7, 227-51
Daudt, H. (1977) 'The Political Future of the Welfare State', *Netherlands Journal of Sociology*, 13, 89-106
Davies, B.P., Barton, A. and McMillan, I. (1972) *Variations Among Children's Services in British Urban Authorities*, Bell, London
Davies, R.B. and O'Farrell, P.N. (1981) 'A Spatial and Temporal Analysis of Second Home Ownership in West Wales', *Geoforum*, 12, 11-178
Davis, N. (1978) 'Population Trends: a European Overview', *Population Trends*, 12, 10-12
Deakin, N. (1970) *Colour, Citizenship and British Society*, Panther, London
Dear, M. and Clark, G. (1978) 'The State and Geographic Process: A Critical Review', *Environment and Planning*, 10A, 173-83
Dearlove, J. (1973) *The Politics of Policy in Local Government*, Cambridge University Press, London
Delaisi, F. (1929) *Les Deux Europes*, Payot, Paris
Della Pergola, G. (1975) *Città e Conflitto Sociale*, Feltrinelli, Milan
Dell'Atti, A. (1979) 'Sullo spopolamento montano in Calabria', *Rassegna Economica*, 43, 1477-88
Del Panta, L. (1979) 'Italy' in W.R. Lee (ed.), *European Demography and Economic Growth*, Croom Helm, London, pp. 196-235
Demangeon, A. (1927) 'La Géographie de l'habitat rural', *Annales de géographie*, 36, 1-17
Dench, G. (1975) *Maltese in London: A Case Study in the Erosion of Ethnic Consciousness*, Routledge & Kegan Paul, London
Dennett, J., James, E., Room, G. and Watson, P. (1982) *Europe Against Poverty*, Bedford Square Press, London
Dente, B., Mayntz, R. and Sharpe, L.J. (1977) *Il Governo Locale in Europa*, Edizioni di Communità, Milan
—— and Regonini, G. (1980) 'Urban Policy and Political Legitimation: the Case of Italian Neighbourhood Councils', *International Political Science Review*, 1, 187-202
Department of the Environment (1975) *A Survey of Second Homes in the South West*, HMSO, London
—— (1977) *Unequal City: Final Report of the Birmingham Inner Area Study*, HMSO, London
Deprez, P. (1979) 'The Low Countries', in W.R. Lee (ed.), *European Demography*

and Economic Development, Croom Helm, London, pp. 236-83

Despicht, N. (1980) 'Centre and Periphery in Europe', in J. De Bandt, P. Mandi and D. Seers (eds), *European Studies in Development*, Macmillan, London, pp. 38-41

Deutsch, K.W. (1966) *Nationalism and Social Communication*, MIT Press, Cambridge, Mass.

——, Edinger, L.J., Macridis, R.C. and Merritt, R.K. (1967) *France, Germany and the Western Alliance: A Study of Elite Attitudes on European Integration and World Politics*, Scribner's, New York

Dicken, P. (1980) 'Foreign Direct Investment in European Manufacturing Industry: the Changing Position of the UK as a Host Country', *Geoforum, 11*, 289-313

—— and Lloyd, P.E. (1980) 'Patterns and Processes of Change in the Spatial Distribution of Foreign-controlled Manufacturing Employment in the UK, 1963-1975', *Environment and Planning, 12A*, 1405-26

—— and —— (1981) *Modern Western Society*, Harper & Row, London

Dickinson, R.E. (1951) *The West European City*, Routledge & Kegan Paul, London

Diederiks, H.A. (1981) 'Patterns of Urban Growth since 1500, Mainly in Western Europe', in H. Schmol (ed.), *Patterns of European Urbanisation Since 1500*, Croom Helm, London, pp. 1-30

Diem, A. (1963) 'An Evaluation of Land Reform and Reclamation in Italy', *Canadian Geographer, 7*, 182-91

—— (1979) *Western Europe: A Geographical Analysis*, Wiley, New York

Dogan, M. (1967) 'Political Cleavage and Social Stratification in France and Italy', in S.M. Lipset and S. Rokkan (eds), *Party Systems and Voter Alignments: Cross-National Perspectives*, Collier-Macmillan, London

Donnison, D. and Soto, P. (1980) *The Good City*, Heinemann, London

Donolo, C. (1980) 'Social Change and Transformation of the State in Italy', in R. Scase (ed.), *The State in Western Europe*, Croom Helm, London, pp. 164-96

Döpp, W. and Leib, J. (1980) 'Gastarbeiter in Stadtallendorf (Hessen), *Zeitschrift fur Bevölkerungswissenschaft, 6*, 59-84

Douglas, W.A. (1971) 'Rural Exodus in Two Spanish Basque Villages: A Cultural Explanation', *American Anthropologist, 73*, 1100-14

Dower, M. (1980) *Jobs in the Countryside*, Rural Advisory Committee, National Council for Voluntary Associations, London

Draaisma, J. and van Hoogstraten, P. (1983) 'The Squatter Movement in Amsterdam', *International Journal of Urban and Regional Research, 7*, 406-16

Drake, M. (1969) *Population and Society in Norway 1735-1865*, Cambridge University Press, Cambridge

Drewer, S. (1974) 'The Economic Impact of Immigrant Workers in Western Europe', *European Studies, 18*, 1-4

Drewett, R. and Rossi, A. (1981) 'General Urbanisation Trends in Western Europe', in L.H. Klaasen, W.T.M. Molle and J.H.P. Paelinck (eds), *Dynamics of Urban Development*, Gower, Aldershot, pp. 119-36

Drudy, P. (1978) 'Depopulation in a Prosperous Agricultural Sub-region', *Regional Studies, 12*, 49-60

Dunford, M., Geddes, M., and Perrons, D. (1981) 'Regional Policy and the Crisis in the UK: a Long-run Perspective', *International Journal of Urban and Regional Research, 5*, 411-26

Durand-Drouhin, J.L. and Szwengrub, L.M. (eds) (1981) *Rural Community Studies in Europe*, volume 1, Pergamon, Oxford

—— and —— (eds) (1982) *Rural Community Studies in Europe*, volume 2, Pergamon, Oxford

Dyer, C. (1978) *Population and Society in Twentieth-Century France*, Hodder & Stoughton, London

Dyson, K.H.F. (1980) *The State Tradition in Western Europe: A Study of an Idea and Institution*, Martin Robertson, Oxford

Easterlin, R.A. (1966) 'Economic and Demographic Interactions: Long Swings – Economic Growth', *American Economic Review*, 56, 1063-104

Edwards, J.A. (1971) 'The Viability of Lower Size-order Settlements in Rural Areas', *Sociologia Ruralis*, 11, 247-76

EFTA (1973) *National Settlement Strategies*, European Free Trade Association, Geneva

Eggeling, W.J. (1981) 'Turkische Gastarbeiter in der Stadt Hattinger', *Wuppertaler Geographische Studien*, 2, 225-40

Eikaas, F.H. (1979) 'You Can't Go Home Again? Culture Shock and Patterns of Adaptation in Norwegian Returnees', *Papers in Anthropology*, 20, 105-15

Einem, E. von (1982) 'National Urban Policy: the Case of West Germany', *Journal of the American Planning Association*, 48, 9-23

Elder, N. (1979) 'The Functions of the Modern State', in J. Hayward and R.N. Berki (eds), *State and Society in Contemporary Europe*, Martin Robertson, London, pp. 58-74

Emerson, A.R. and Crompton, R. (1968) *Suffolk: Some Social Trends* (mimeo), Suffolk Rural Community Council

Entzinger, H. (1978) *Return Migration from Western European to Mediterranean Countries*, ILO World Employment Programme, ILO-WEP 2-26/WP23, Geneva

Erlandsson, U. (1979) 'Contact Potentials in the European System of Cities', in H. Folmer and J. Oosterhaven (eds), *Spatial Inequalities and Regional Development*, Martinus Nijhoff, Boston, pp. 93-116

Ermisch, J. (1982a) 'Investigations into the Causes of the Postwar Fertility Swings', in D. Eversley and W. Kollmann (eds), *Population Change and Social Planning*, Arnold, London, pp. 141-55

—— (1982b) 'The Labour Market: Historical Development and Hypotheses', in D. Eversley and W. Kollmann (eds), *Population Change and Social Planning*, Arnold, London, pp. 156-209

Estrin, S. and Holmes, P. (1983) *French Planning in Theory and Practice*, Allen & Unwin, Hemel Hempstead

Euromonitor (1977) *Consumer Europe 1977*, Euromonitor Publications, London

European Parliament (1981) *Europe Today. State of European Integration, 1980-81*, European Parliament Secretariat, Luxembourg

Eurostat (1982) *Demographic Statistics 1980*, Commission of the European Communities, European Statistics, Luxembourg

Evans, R.H. (1981) 'Local Government Reform in Italy, 1945-1979', in A. Gunlicks (ed.), *Local Government Reform and Reorganisation*, Kennikat Press, Port Washington, NY, pp. 112-30

Eversley, D. (1978) 'Welfare' in Council of Europe, *Population Decline in Europe*, Arnold, London, pp. 115-42

—— (1982) 'Social Policy: Implications of Changes in the Demographic Situation', in D. Eversley and W. Kollmann (eds), *Population Change and Social Planning*, Arnold, London, pp. 374-413

Ewers, H.J. and Wettmann, R.W. (1980) 'Innovation-oriented Regional Policy', *Regional Studies*, 14, 161-79

Feichtinger, G. (1975) *Are Economically Dependent Groups Likely to Become a Significantly Larger Proportion of the Population as a Whole?* Council of Europe, Seminar Paper AS/PR/Coll.75(2), Strasbourg

Feld, W.J. and Wildgen, J.K. (1976) *Domestic Political Realities and European Unification: A Study of Mass Publics and Elites in European Community*

Countries, Westview Press, Boulder, Colorado

Festy, P. (1980) 'On the New Context of Marriage in Western Europe', *Population and Development Review, 6*, 311-15

Fielding, A.J. (1982) 'Counterurbanisation in Western Europe', *Progress in Planning, 17*, Pergamon, Oxford

Finer, S.E. (1975) 'State- and Nation-building in Europe: the Role of the Military', in C. Tilly (ed.), *The Formation of National States in Europe*, Princeton University Press, Princeton, pp. 84-163

Firn, J. (1975) 'External Control and Regional Development', *Environment and Planning, 7A*, 393-414

Fishman, J. (1972) *Language and Nationalism: Two Integrative Essays*, Newbury House, Rowley, Mass.

Fitzpatrick, D. (1978) 'The Geography of Irish Nationalism', *Past and Present, 78*, 113-45

Flatrès, P. (1974) 'Problèmes d'aménagement dans les régions d'habitat dispersé de l'Europe de l'Ouest', *Geographia Polonica, 29*, 43-54

Flora, P. and Alber, J. (1981) 'Modernisation, Democratisation and the Development of Welfare States in Western Europe', in P. Flora and A.J. Heidenheimer (eds), *The Development of Welfare States in Europe and America*, Transaction Books, New Brunswick, NJ, pp. 37-80

Folmer, H. and Oosterhaven, J. (eds) (1979) *Spatial Inequalities and Regional Development*, Martinus Nijhoff, The Hague

Ford, N.J. (1982) 'Consciousness and Lifestyle: Alternative Developments in the Culture and Economy of Rural Dyfed', University of Wales Doctoral Dissertation, Aberystwyth

Forsyth, D. (1980) 'Urban Incomers and Community Change: the Impact of Migrants from the City on Life in an Orkney Community', *Sociologia Ruralis, 20*, 287-307

—— (1982) 'Urban-rural Migration and the Pastoral Ideal: an Orkney Case', in A. Jackson (ed.), *Way of Life: Integration and Immigration*, SSRC North Sea Oil Panel, Occasional Paper no. 12, 22-45

—— (1984) 'The Social Effect of Primary School Closure', in A. Bradley and P. Lowe (eds), *Locality and Rurality: Economy and Society in Rural Regions*, Geo Books, Norwich

Fothergill, S. and Gudgin, G. (1982) *Unequal Growth: Urban and Regional Employment Change in the UK*, Heinemann, London

—— and —— (1983) 'Trends in Regional Manufacturing Employment: the Main Influences', in J.B. Goddard and A.G. Champion (eds), *The Urban and Regional Transformation of Britain*, Methuen, London, pp. 27-50

Frank, W. (1983) 'Part-time Farming, Underemployment and Double Activity of Farmers in the EEC', *Sociologia Ruralis, 23*, 20-7

Franklin, S.H. (1969) *The European Peasantry: The Final Phase*, Methuen, London

—— (1971) *Rural Societies*, Macmillan, London

Frears, J. (1978) 'The French National Assembly Elections of 1978', *Government and Opposition, 13*, 323-40

Fridlizius, G. (1979) 'Sweden', in W.R. Lee (ed.), *European Demography and Economic Growth*, Croom Helm, London, pp. 340-405

Fried, R.C. (1971) 'Communism, Urban Budgets and the Two Italies: A Case Study in Comparative Urban Governments', *Journal of Politics, 33*, 1008-51

—— (1973) *Planning the Eternal City: Roman Politics and Planning Since World War 2*, Yale University Press, New Haven

—— (1974) 'Politics, Economics and Federalism: Aspects of Urban Government in Mittel-Europa', in T.N. Clark (ed.), *European Communities*, Sage, Beverly Hills

—— (1976) 'Party and Policy in West German Cities', *American Political Science Review*, 70, 11-24
Friis, P. (1980) 'Regional Problems in Denmark: Myth or Reality?' *Dunelm Translations, 4*, Department of Geography, University of Durham
Frobel, F., Heinrichs, J. and Kreye, O. (1980) *The New International Division of Labour*, Cambridge University Press, Cambridge
Fuerst, J.S. (ed.) (1974) *Public Housing in Europe and America*, Croom Helm, London
Galbraith, J.K. (1967) *The New Industrial State*, Signet, New York
—— (1975) *Economics and the Public Purpose*, Penguin, Harmondsworth
Ganser, K. (1982) 'Eine Ökonomische und Ökologische Perspective für die Agrar-politik', *Geographische Rundschau, 34*, 82-7
Gaunt, D. (1976) 'Familj, Lushall och arbetsintensitet', *Scandia, 42*, 32-59
Geddes, M. (1979) *Uneven Development and the Scottish Highlands*, University of Sussex, Department of Urban and Regional Studies, Working Paper No. 17
George, V. and Lawson, R. (eds) (1980) *Poverty and Inequality in Common Market Countries*, Routledge & Kegan Paul, London
Gerschenkron, A. (1966) *Economic Backwardness in Historical Perspective*, Praeger, Cambridge, Mass.
Gilkey, G.R. (1968) 'Migration and Repatriation', in F.D. Scott (ed.), *World Migration in Modern Times*, Prentice-Hall, Englewood Cliffs, pp. 42-54
Giner, S. (1982) 'Political Economy, Legitimation and the State in Southern Europe', *British Journal of Sociology, 33*, 172-99
—— and Archer, M.S. (1978) *Contemporary Europe: Social Structures and Cultural Patterns*, Routledge & Kegan Paul, London
—— and Salcedo, J. (1978) 'Migrant Workers in European Social Structures', in S. Giner and M.S. Archer (eds), *Contemporary Europe: Social Structures and Cultural Patterns*, Routledge & Kegan Paul, London, pp. 94-123
—— and Sevila-Guzman, E. (1980) 'The Demise of the Peasant', *Sociologia Ruralis*, 22, 13-27
Ginsburgh, V. and Pestieau, P. (1981) 'Local Government Expenditure in Belgium: do Political Distinctions Matter?', *European Journal of Political Research*, 9, 169-80
Glass, D. (1967) *Population Policies and Movements in Europe*, Cass, London
Glennerster, H. (1980) 'Prime Cuts: Public Expenditure and Social Service Planning in a Hostile Environment', *Policy and Politics, 8*
Gloag, J. (1979) 'Nazi Interlude', *Architectural Review, 116*, 326-8
Gmelch, G. (1980) 'Return Migration', *Annual Review of Anthropology*, 9, 135-59
Gober-Meyers, P. (1978) 'Employment-motivated Migration and Economic Growth in Post-industrial Market Economies', *Progress in Human Geography*, 2, 207-29
Goddard, J.B. (1983) 'Structural Change in the British Space-economy', in J.B. Goddard and A.G. Champion (eds), *The Urban and Regional Transformation of Britain*, Methuen, London, pp. 1-26
Goldfield, D. (1982) 'National Urban Policy in Sweden', *Journal of the American Planning Association, 48*, 24-38
Goldschmidt, W. (1978) 'Large-scale Farming and the Rural Social Structure', *Rural Sociology, 43*, 362-6
Gonen, A. (1981) 'Tourism and Coastal Settlement Processes in the Mediterranean Region', *Ekistics, 48*, 378-81
Goodyear, P.M. and Eastwood, D.A. (1978) 'Spatial Variations in Levels of Living in Northern Ireland', *Irish Geography, 11*, 54-67
Gottmann, J. (1969) *A Geography of Europe*, 4th edn., Holt, Rinehart & Winston,

London
—— (ed.) (1980) *Centre and Periphery*, Sage, London
—— (1980) 'Confronting Centre and Periphery', in J. Gottmann (ed.), *Centre and Periphery*, Sage, London, pp. 11-26
Gough, I. (1982) 'The Crisis of the British Welfare State', in N. Fainstein and S. Fainstein (eds), *Urban Policy Under Capitalism*, Sage, Beverly Hills, pp. 43-64
Gourevitch, P.A. (1980) *Paris and the Provinces*, Allen & Unwin, London
Granotier, B. (1970) *Les travailleurs immigrés en France*, Maspero, Paris
Gras, S. (1982) 'Regionalism and Autonomy in Alsace Since 1918', in S. Rokkan and D. Urwin (eds), *The Politics of Territorial Identity*, Sage, London, pp. 309-54
Grasmuck, S. (1980) 'Ideology of Ethnoregionalism: the Case of Scotland', *Politics and Society*, *9*, 471-86
Gravier, J.F. (1947) *Paris et le désert français*, Flammarion, Paris
Graziano, L. (1978) 'Center-periphery Relations and the Italian Crisis: the Problem of Clientelism', in S. Tarrow, P.J. Katzenstein and L. Graziano (eds), *Territorial Politics in Industrial Nations*, Praeger, New York, pp. 290-326
Green, D., Haselgrove, C. and Spriggs, M. (eds) (1978) *Social Organisation and Settlement*, British Archaeological Reports, International Series, 47, (2 vols.)
Green, P. (1964) 'Drymen: Village Growth and Community Problems', *Sociological Review*, *4*, 52-62
Greenwood, D. (1972) 'Tourism as an Agent of Change: A Spanish Basque Case', *Ethnology*, *11*, 80-91
Griffith, G.T. (1926) *Population Problems of the Age of Malthus*, Cass, London
Grillo, R.D. (1980) 'Introduction', in R.D. Grillo (ed.), *'Nation' and 'State' in Europe: Anthropological Perspectives*, Academic Press, London, pp. 1-30
Guilmot, P. (1978) 'The Demographic Background', in Council of Europe, *Population Decline in Europe*, Arnold, London, pp. 3-49
Gunlicks, A.B. (ed.) (1981) *Local Government Reform and Reorganisation*, Kennikat Press, Port Washington, New York
Gustaffsson, M. (1979) 'Migration Research and Theories with Special Reference to Western Europe', *Siirtolaisuns*, *6*, 18-29
Gutkind, E.A. (1965) *Urban Development in the Alpine and Scandinavian Countries*, Collier-Macmillan, London
—— (1967) *Urban Development in Southern Europe: Spain and Portugal*, Collier-Macmillan, London
—— (1969) *Urban Development in Southern Europe: Italy and Greece*, Collier-Macmillan, London
—— (1970) *Urban Development in Western Europe: France and Belgium*, Collier-Macmillan, London
—— (1971) *Urban Development in Western Europe: Great Britain and the Netherlands*, Collier-Macmillan, London
Haan, E. de (1976) 'Foreign Workers and Social Sources in Federal Germany', in N. Abadan-Unat (ed.), *Turkish Workers in Europe 1960-1975*, Brill, Leiden, pp. 346-62
Haavo-Manila, E. and Kari, K. (1979) 'Demographic Background of Changes in the Life Patterns of Families in the Nordic Countries', *Working Paper*, *11*, Department of Sociology, University of Helsinki, Helsinki
Haegen, H. van der (ed.) (1983) *West European Settlement Systems*, Instituut voor Economische en Sociale Geografie, Katholicke Universiteit, Leuven, Belgium
Hain, P. (1976) 'Neighbourhood Councils: the Attitudes of the Central Authorities', *Community Development Journal*, *9*, 2-9
Hajdu, J. (1983) 'Postwar Development and Planning of West German Cities', in

T. Wild (ed.), *Urban and Regional Change in West Germany*, Croom Helm, London, pp. 16-39

Hajnal, J. (1965) 'European Marriage Patterns in Perspective', in D. Glass and D. Eversley (eds), *Population History*, Arnold, London, pp. 101-43

Hall, P. (1969) *Europe 2000*, Faber & Faber, London

—— (1977) *The World Cities* (2nd edition), Weidenfeld & Nicolson, London

—— (1981a) 'The Geography of the Fifth Kondratieff Cycle', *New Society*, 535-7

—— (1981b) 'Regional Development in the EEC: A Look Back and a Look Forward', *Built Environment*, 7, 229-32

—— (1982) *Urban and Regional Planning* (2nd edition), Penguin, Harmondsworth

—— and Hay, D. (1980) *Growth Centres in the European Urban System*, Heinemann, London

Hamilton, F.E.I. (1976) 'Multinational Enterprise and the European Economic Community', *Tijdschrift voor Economische en Sociale Geografie*, 67, 258-77

Handley, D.H. (1981) 'Public Opinion and European Integration', *European Journal of Political Research*, 9, 335-64

Handlin, O. (1956) 'Immigrants Who Go Back', *Atlantic*, 198, 70-4

Hansen, E.J. and Geckler, S. (1978) 'The Danish Welfare Survey', *Acta Sociologica*, 21, 385-8

Hansen, T. (1981) 'The Dynamics of Local Expenditure Growth: Local Government Finance in Sweden', in L.J. Sharpe (ed.), *The Local Fiscal Crisis in Western Europe*, Sage, Beverly Hills, pp. 165-94

Harman, R. (1982) 'Rural Services: Change in N E Norfolk', *Policy and Politics*, 10, 1982, pp. 477-94

Harvie, C. (1977) *Scotland and Nationalism*, Allen & Unwin, London

Hechter, M. (1975) *Internal Colonialism: the Celtic Fringe in British National Development*, Routledge & Kegan Paul, London

—— and Levi, M. (1979) 'The Comparative Analysis of Ethnoregional Movements', *Ethnic and Racial Studies*, 2, 206-74

Heclo, H. (1981) 'Toward a New Welfare State?', in P. Flora and A.J. Heidenheimer (eds), *The Development of Welfare States in Europe and America*, Transaction Books, New Brunswick, NJ, pp. 383-406

Heiberg, M. (1982) 'Urban Politics and Rural Culture', in S. Rokkan and D. Urwin (eds), *The Politics of Territorial Identity*, Sage, London, pp. 355-88

Heisler, M.O. (1977) 'Managing Ethnic Conflict in Belgium', *Annals, American Academy of Political and Social Science*, 433, 32-46

Hepburn, A.C. (1980) *The Conflict of Nationality in Northern Ireland*, Arnold, London

Herfurth, M. and Hogweg-de-Haart, M. (eds) (1982) *Social Integration of Migrant Workers and other Ethnic Minorities: A Documentation of Current Research*, Pergamon, Oxford

Hervo, M. and Charras, M. (1971) *Bidonvilles*, Maspero, Paris

Hiro, D. (1973) *Black British, White British*, Penguin, Harmondsworth

Hirsch, F. (1977) *The Social Limits to Growth*, Routledge & Kegan Paul, London

Hirsch, G.P. and Maunder, A.H. (1978) *Farm Amalgamation in Western Europe*, Saxon House, Farnborough

Hirschmann, A.O. (1958) *The Strategy of Economic Development*, Yale University Press, New Haven

Hodge, I. and Whitby, M. (1981) *Rural Employment: Trends, Options, Choices*, Methuen, London

Hoffman, G.W. (1977) 'Regional Policies and Regional Consciousness in Europe's Multinational Societies', *Geoforum*, 8, 121-9

Hoffman, S. (1982) 'Reflections on the Nation-State in Western Europe Today',

Journal of Common Market Studies, 21, 21-38

Hofsten, E. and Lundström, H. (1978) 'Swedish Population History', *Urval, 8*, Skriftserie utgiven uv statistika centralbyrån, Stockholm

Höhn, C. (1981) 'Les différences internationales de mortalité infantile: illusion ou réalité?', *Population, 36*, 791-816

Holland, S. (1976a) 'Meso-economies, Multinational Capital and Regional Inequality', in R. Lee and P. Ogden, (eds), *Economy and Society in the EEC: Spatial Perspectives*, Saxon House, Farnborough, pp. 38-62

—— (1976b) *Capital versus the Regions*, Macmillan, London

—— (1980) *UnCommon Market*, Macmillan, London

Hooson, D.J.M. (1960) 'The Distribution of Population as the Essential Geographical Expression', *Canadian Geographer, 17*, 10-20

House, J.W. (ed.) (1982) *The UK Space* (3rd edition), Weidenfeld & Nicolson, London

Houston, J.M. (1963) *A Social Geography of Europe*, Duckworth, London

Hudson, R. (1983a) 'Capital Accumulation and Regional Problems: a Study of North-east England', in F.E.I. Hamilton and G. Linge (eds), *Regional Industrial Systems*, Wiley, Chichester

—— (1983b) 'Regional Labour Reserves and Industrialisation in the EEC', *Area, 15*, 223-30

—— and Lewis, J.R. (eds) (1982) *Regional Planning in Europe*, Pion, London

Hughes, R. (1980) 'The Shock of the New', *The Listener*, 389-91, 425-7, 495-9

Hume, I.M. (1973) 'Migrant Workers in Europe', *Finance and Development, 10*, 2-6

Ilbery, B.W. (1981) *Western Europe: A Systematic Human Geography*, Oxford University Press, Oxford

Inglehart, R. (1970a) 'Cognitive Mobilisation and European Integration', *Comparative Politics, 5*, 112-21

—— (1970b) 'Public Opinion and Regional Integration', *International Organisation, 24*, 764-95

—— (1971) 'The Silent Revolution in Europe: Intergenerational Change in Industrial Societies', *American Political Science Review, 65*, 991-1017

—— (1977) *The Silent Revolution: Changing Values and Political Styles among Western Publics*, Princeton University Press, Princeton

International Labour Office (1974) *Some Growing Employment Problems in Europe*, ILO, Geneva

Jackman, R.W. (1980) 'Socialist Parties and Income Inequality in Western Industrial Societies', *Journal of Politics, 42*, 135-49

Jacob, J.E. (1975) 'The Basques of France: a Case of Peripheral Ethnonationalism', *Political Anthropology, 1*, 67-87

Jacoby, E.H. (1959) *Land Consolidation in Europe*, International Institute for Land Reclamation and Improvement, Wageningen

Jain, S. (1975) *Size Distribution of Income*, The World Bank, Washington, DC

Jensen, J. and Simonsen, K. (1981) 'The Local State, Planning and Social Movements', *Acta Sociologica, 24*, 279-91

Johansson, S. (1973) 'The Level of Living Survey: a presentation', *Acta Sociologica, 16*, 211-19

Johnson, R.W. (1981) *The Long March of the French Left*, Macmillan, London

Johnston, R.J. (1977) 'National Sovereignty and National Power in European Institutions', *Environment and Planning, 9A*, 569-77

—— (1979a) *Political, Electoral and Spatial Systems*, Oxford University Press, London

—— (1979b) 'The Spatial Impact of Fiscal Change in Britain: Regional Policy in Reverse?', *Environment and Planning, 11A*, 1439-44

—— (1982) *Geography and the State*, Macmillan, London
——, O'Neill, A.B. and Taylor, P.J. (1983) 'The Changing Electoral Geography of the Netherlands', *Tijdschrift voor Economische en Sociale Geografie, 74*, 185-95
Jones, E.L. (1981) *The European Miracle*, Cambridge University Press, Cambridge
Jones, H.R. (1965) 'A Study of Rural Migration in Central Wales', *Transactions, Institute of British Geographers, 37*, 31-45
—— (1981) *A Population Geography*, Harper & Row, London
Jones, P.N. (1967) *The Segregation of Immigrant Communities in the City of Birmingham, 1961*, University of Hull, Hull
—— (1979) 'Ethnic Areas in British Cities', in D. Herbert and D. Smith (eds), *Social Problems and the City*, Oxford University Press, Oxford, pp. 158-85
—— (1983) 'Ethnic Population Succession in a West German City 1974-1980: the Case of Nuremburg', *Geography, 68*, 121-32
Jordan, T.G. (1973) *The European Culture Area*, Harper & Row, London
Kaa, D.J. van de (1980) 'Recent Trends in Fertility in Western Europe', in R.W. Hiorns (ed.), *Demographic Patterns in Developed Societies*, Taylor & Francis, London, pp. 55-82
Kain, R. (ed.) (1980) *Planning for Conservation*, Mansell, London
Kalk, E. (ed.) (1971) *Regional Planning and Regional Government in Europe*, International Union of Local Authorities, The Hague
Kane, T.T. (1978) 'Social Problems and Ethnic Change: Europe's "Guest Workers" ', *Intercom, 6*, 7-9
Kariel, H.G. and Kariel, P.E. (1982) 'Socio-cultural Impacts of Tourism: An Example from the Austrian Alps', *Geografiska Annaler*, 1-16
Karn, V.A. (1977) *Retiring to the Seaside*, Routledge & Kegan Paul, London
Kayser, B. (1967) 'The Situation of the Returning Migrant on the Labour Market in Greece', in OECD, *Emigrant Workers Returning to their Home Country*, OECD, Paris, pp. 169-76
—— (1977) 'The Effects of International Migration on the Geographical Distribution of Population in Europe', *Population Studies, 2*, Council of Europe, Strasbourg
Kearsley, G.W. and Srivastava, S.R. (1974) 'The Spatial Evolution of Glasgow's Asian Community', *Scottish Geographical Magazine, 90*, 110-24
Keeble, D.E. (1976) *Industrial Location and Planning in the United Kingdom*, Methuen, London
—— (1980) 'Industrial Decline, Regional Policy and the Urban-rural Manufacturing Shift in the UK', *Environment and Planning, 12A*, 945-62
——, Owens, P.L. and Thompson, C. (1981) 'EEC Regional Disparities and Trends in the 1970s', *Built Environment, 7*, 154-61
——, Owens, P.L. and Thompson, C. (1982) 'Regional Accessibility and Economic Potential in the European Community', *Regional Studies, 16*, 419-31
Kemper, N.J. and De Smidt, M. (1980) 'Foreign Manufacturing Establishments in the Netherlands', *Tijdschrift voor Economische en Sociale Geografie, 71*, 21-40
Kennedy-Brenner, C. (1979) *Foreign Workers and Immigration Policy: the Case of France*, OECD, Paris
Kenny, M. (1972) 'The Return of the Spanish Emigrant', *Nord Nytt, 2*, 119-29
Keyfitz, N. (1973) 'Individual Mobility in a Stationary Population', *Population Studies, 27*, 210-27
Kiljunen, M-L. (1980) 'Regional Disparities and Policy in the EEC', in D. Seers and C. Vaitsos (eds), *Integration and Unequal Development*, Macmillan, London, pp. 199-222
King, R.L. (1971) 'Italian Land Reform: Critique, Effects, Evaluation', *Tijdschrift*

voor Economische en Sociale Geografie, *62*, 368-82
— (1973) *Land Reform: The Italian Experience*, Bell, London
— (1976) 'The Evolution of International Labour Migration Movements Concerning the EEC', *Tijdschrift voor Economische en Sociale Geografie*, *67*, 66-82
— (1977) 'Problems of Return Migration: a Case Study of Italians Returning from Britain', *Tijdschrift voor Economische en Sociale Geografie*, *68*, 241-5
— (1978) 'Return Migration: a Review of Some Cases from Southern Europe', *Mediterranean Studies*, *1*, 112-26
— (1982) 'Southern Europe: Dependency or Development?', *Geography*, *67*, 221-34
— and — (1983) 'Structural Change in Agriculture: the Geography of Land Consolidation', *Progress in Human Geography*, *7*, 471-501
— and Burton, S. (1983) 'Structural Change in Agriculture: the Geography of Land Consolidation', *Progress in Human Geography*, *7*, 471-501
Kirby, A. (1979) *Education, Health and Housing*, Saxon House, Farnborough
— (1982a) *The Politics of Location*, Methuen, London
— (1982b) 'The External Relations of the Local State in Britain: Some Examples', in K.R. Cox and R.J. Johnston (eds), *Conflict, Politics and the Urban Scene*, Longman, London, pp. 88-104
Kirk, M. (1981) *Demographic and Social Change in Europe: 1975-2000*, Liverpool University Press, Liverpool
Kitzinger, U. (1967) *European Common Market and Community*, Routledge & Kegan Paul, London
Kjellberg, F. (1979) 'A Comparative View of Municipal Decentralisation: Neighbourhood Democracy in Oslo and Bologna', in L.J. Sharpe (ed.), *Decentralist Trends in Western Democracies*, Sage, London, pp. 81-118
— (1981) 'The Expansion and Standardisation of Local Finance in Norway', in L.J. Sharpe (ed.), *The Local Fiscal Crisis in Western Europe*, Sage, Beverly Hills, pp. 125-64
Klaasen, L.H., Molle, W.T.M. and Paelinck, J.H.P. (eds) (1981), *Dynamics of Urban Development*, Gower, Aldershot
Klein, L. (1981) 'The European Community's Regional Policy', *Built Environment*, *7*, 182-9
Knight, D.B. (1982) 'Identity and Territory: Geographical Perspectives on Nationalism and Regionalism', *Annals, Association of American Geographers*, *72*, 514-31
Knodel, J. (1974) *The Decline of Fertility in Germany 1871-1939*, Princeton University Press, Princeton
— and Walle, E. van de (1979) 'Lessons from the Past: Policy Implications of Historical Fertility Studies', *Population and Development Review*, *5*, 217-46
Knowles, R.D. (1981) 'Malapportionment in Norway's Parliamentary Elections Since 1921', *Norsk Geografisk Tidsskrift*, *35*, 147-59
Knox, P.L. (1974) 'Spatial Variations in Level of Living in England and Wales in 1961', *Transactions, Institute of British Geographers*, *62*, 1-24
— (1975) *Social Well-Being: A Spatial Perspective*, Clarendon Press, Oxford
— (1978) 'Community Councils, Electoral Districts and Social Geography', *Area*, *10*, 387-91
— (1982a) 'Regional Inequality and the Welfare State: Convergence and Divergence in Levels of Living in the United Kingdom 1951-1971', *Social Indicators Research*, *10*, 319-35
— (1982b) *Urban Social Geography*, Longman, London
— (1982c) 'The Social Production of the Built Environment', *Ekistics*, *49*, 291-7

—— and Cottam, M.B. (1981a) 'A Welfare Approach to Rural Geography: Contrasting Perspectives on the Quality of Highland Life', *Transactions, Institute of British Geographers, 6*, 433-50

—— and —— (1981b) 'Rural Deprivation in Scotland: A Preliminary Assessment', *Tijdschrift voor Economische en Sociale Geografie, 72*, 162-75

—— and MacLaran, A. (1978) 'Values and Perceptions in Descriptive Approaches to Urban Social Geography', in D. Herbert and R.J. Johnston (eds), *Geography and the Urban Environment*, vol. 1, Wiley, Chichester, pp. 197-248

—— and Scarth, A. (1977) 'The Quality of Life in France', *Geography, 62*, 9-16

—— and Kirby, A. (1984) 'Public Provision and Quality of Life', in A. Kirby and J. Short (eds), *Britain Now: A Contemporary Human Geography*, Macmillan, London

Koch, H. (1979) 'Remsfeld – Lebensverhältnisse in einem Dorf Nordhessens', in O. Poppinga (ed.), *Produktion und Lebensverhältnisse auf dem Land*, Leviathan, Stuttgart, pp. 215-35

Koch, R. (1980) ' "Counterurbanisation" auch in Westeuropa?', *Informationen zür Raumentwicklung, 2*, 59-69

Kolinsky, M. (ed.) (1978) *Divided Loyalties*, Manchester University Press, Manchester

Kondratieff, N.D. (1935) 'The Long Waves in Economic Life', *Review of Economics and Statistics, 17*

Kosiński, L.A. (1970) *The Population of Europe*, Longman, London

Kraus, F. (1981) 'The Historical Development of Income Inequality in Western Europe and the United States', in P. Flora and A.J. Heidenheimer (eds), *The Development of Welfare States in Europe and America*, Transaction Books, New Brunswick, NJ, pp. 187-236

Krejci, J. (1978) 'Ethnic Problems in Europe', in S. Giner and M.S. Archer (eds), *Contemporary Europe: Social Structures and Cultural Patterns*, Routledge & Kegan Paul, London, pp. 124-71

Kuklinski, A. (ed.) (1981) *Polarized Development and Regional Policies*, Mouton, Paris

Künnecke, B.H. (1974) 'Sozialbrache – A Phenomenon in the Rural Landscape of Germany', *Professional Geographer, 26*, 412-15

Langdalen, E. (1980) 'Second Homes in Norway: a Controversial Planning Problem', *Norsk Geografisk Tijdsskrift, 34*, 139-44

Langer, W. (1975) 'American Foods and Europe's Population Growth 1750-1850', *Journal of Social History, 5*, 51-66

Law, C.M. (1980) 'The Foreign Company's Location Investment Decision and its Role in British Regional Development', *Tijdschrift voor Economische en Sociale Geografie, 71*, 15-20

—— and Warnes, A.M. (1980) 'The Characteristics of Retired Migrants', *Geography and the Urban Environment, 3*, 175-222

—— and —— (1981) 'The Destination Decision in Retirement Migration', in A.M. Warnes (ed.), *Geographical Perspectives on the Elderly*, Wiley, Chichester

Lawson, R. (1980) 'Poverty and Inequality in West Germany', in V. George and R. Lawson (eds), *Poverty and Inequality in Common Market Countries*, Routledge & Kegan Paul, London, pp. 195-232

—— and George, V. (1980) 'An Assessment', in V. George and R. Lawson (eds), *Poverty and Inequality in Common Market Countries*, Routledge & Kegan Paul, London, pp. 233-42

Leach, G. (1975) *Energy and Food Production*, IPC Science and Technology Press, London

Le Bras, H. (1974) 'Le mythe de la population stationnaire', *Prospectives, 3*, 71-82

Lee, R. (1976) 'Integration, Spatial Structure and the Capitalist Mode of Production', in R. Lee and P. Ogden (eds), *Economy and Society in the EEC*, Saxon House, Farnborough, pp. 11-37
–– and Ogden, P. (eds) (1976) *Economy and Society in the EEC*, Saxon House, Farnborough
Lee, T.R. (1977) *Race and Residence*, Clarendon Press, Oxford
Lee, W.R. (ed.) (1979a) *European Demography and Economic Growth*, Croom Helm, London
–– (1979b) 'Population Growth, Economic Development and Social Change in Europe 1750-1970', in W.R. Lee (ed.), *European Demography and Economic Growth*, Croom Helm, London, pp. 10-26
Leeson, G.W. (1981) 'The Elderly in Denmark in 1980: Consequences of a Mortality Decline', *European Demographic Information Bulletin*, *12*, 89-100
Lefèvre, M.A. (1926) *L'Habitat rural en Belgique*, Loew, Liège
Leitner, H. (1981) 'Struktur und Determinanten der raümlichen Wohnsegregation der Gastarbeiter in Wien', *Mitteilungen der Österreichischen Geographische Gesellschaft*, *123*, 67-91
Lesthaeghe, R. (1983) 'A Century of Demographic and Cultural Change in Western Europe: an Exploration of Underlying Dimensions', *Population and Development Review*, *9*, 411-36
Lewis, G.J. (1967) 'Commuting and the Village in Mid-Wales', *Geography*, *52*, 294-304
–– (1979) *Rural Communities*, David and Charles, London
Lewis, J. and Hudson, R. (eds) (1984) *Dependent Development in Southern Europe*, Methuen, London
Lianos, T.P. (1975) 'Flows of Greek Outmigration and Return Migration', *International Migration*, *13*, 119-33
Lichtenberger, E. (1972) 'Die Europaische Stadt – Wesen Modelle Probleme', *Raumforsch. Raumord.*, *16*, 3-25
–– (1976) 'The Changing Nature of European Urbanisation', in B.J.L. Berry (ed.), *Urbanization and Counterurbanization*, Urban Affairs Annual Reviews, 11, Sage, Beverly Hills, pp. 81-107
–– (1977) *Die Wiener Altstadt: von der mittelalterlichen Bürgerstadt zur City*, Franz Deuticke, Vienna
–– (1982) 'Guest Workers – A Life in Two Worlds', *Mitteilungen der Österreichischen Geographischen Gesellschaft*, *124*, 28-65
Lijphart, A. (1971) 'Class Voting and Religious Voting in European Democracies', *Acta Politica*, *6*, 158-71
Limouzin, P. (1975) 'Analyse du dynamisme communal en milieu rural', *L'Espace géographique*, *4*, 251-8
Lindberg, L.N. and Scheingold, S.A. (1980) *Europe's Would-Be Polity: Patterns of Change in the European Community*, Prentice-Hall, Englewood Cliffs, New Jersey
Linz, J.J. (1973) 'Early State-building and Late Peripheral Nationalisms Against the State: the Case of Spain', in S.N. Eisenstadt and S. Rokkan (eds), *Building States and Nations*, vol. 2, Sage, Beverly Hills, pp. 32-116
Lipset, S.M. and Rokkan, S. (eds) (1967) *Party Systems and Voter Alignments: Cross-National Perspectives*, Collier-Macmillan, London
Lodge, J. and Herman, V. (1982) *Direct Elections to the European Parliament*, Macmillan, London
Logue, J. (1979) 'The Welfare State: Victim of its Success', *Daedalus*, *108*, 69-87
Lonie, A.A. and Begg, H.M. (1979) 'Comment: Further evidence on the Quest for an Effective Regional Policy, 1934-1937', *Regional Studies*, *13*, 497-500
Lopes, A.S. (1981) 'Regional Development in Portugal', *Built Environment*, *7*, 255-63

Lopreato, J. (1967) *Peasants No More*, Chandler, San Francisco

Lowder, S. (1981) *The Evolution and Identity of Urban Social Areas: the Case of Barcelona*, Occasional Paper no. 4, Department of Geography, University of Glasgow

Lowe, P. and Goyder, J. (eds) (1983) *Environmental Groups in Politics*, Allen & Unwin, London

Lunden, K. (1976) 'Potetdyrkinga og den raskare folketalsvoksteren in Noreg fra 1815', *Historisk Tidsskrift, 55*, 285 (English summary)

Lutz, V. (1962) *Italy: A Study in Economic Development*, Oxford University Press, London

McCleery, A. (1979) 'Rural Depopulation', in *Planning Exchange, Forum Report 15*, Population and Employment in Rural Areas, Glasgow, pp. 9-17

McCrone, G. (1969) *Regional Policy in Britain*, Allen & Unwin, London

McDermott, P.J. (1979) 'Multinational Manufacturing Firms and Regional Development: External Control in the Scottish Electronics Industry', *Scottish Journal of Political Economy, 26*, 287-306

McDonald, J.R. (1969) 'Labour Immigration into France, 1946-1965', *Annals, Association of American Geographers, 59*, 116-34

McDonald, J.S. and McDonald, L. (1982) *The Demography of Aging*, Croom Helm, London

McEntire, D. and Agostini, D. (eds) (1970) *Towards Modern Land Policies*, University of Padua Press, Padua

McKay, J. (1976) *Tramways and Trolleys: The Rise of Urban Mass Transport in Europe*, Princeton University Press, Princeton, NJ

McKennell, A.C. and Andrews, F.M. (1980) 'Models of Cognition and Affect in Perceptions of Well-being', *Social Indicators Research, 8*, 257-98

McKeown, T. (1976) *The Modern Rise of Population*, Arnold, London

——, Brown, R. and Record, R. (1972) 'An Interpretation of the Modern Rise of Population in Europe', *Population Studies, 26*, 345-82

—— and Record, R. (1962) 'Medical Evidence Related to English Population Changes in the Eighteenth Century', *Population Studies, 9*, 119-41

McNabb, R. (1980) 'Segmented Labour Markets, Female Employment and Poverty in Wales', in G. Rees and T.L. Rees (eds), *Poverty and Social Inequality in Wales*, Croom Helm, London, pp. 156-67

Mackay, G. and Laing, G. (1982) *Consumer Problems in Rural Areas*, Scottish Consumer Council, Edinburgh

MacKay, R.R. (1976) 'The Impact of the Regional Employment Premium', in A. Whiting (ed.), *The Economics of Industrial Subsidies*, HMSO, London, pp. 225-42

Mackenzie, R.T. and Silver, A. (1968) *Angels in Marble*, Heinemann, London

MacLaran, A. (1981) 'Area-based Positive Discrimination and the Distribution of Well-being', *Transactions, Institute of British Geographers, 6*, 53-67

Madge, C. and Willmott, P. (1981) *Inner City in Paris and London*, Routledge & Kegan Paul, London

Magnusson, W. (1979) 'The New Neighbourhood Democracy: Anglo-American Experience in Historical Perspective', in L.J. Sharpe (ed.), *Decentralist Trends in Western Societies*, Sage, London, pp. 119-56

Maguire, M. (1983) 'Is There Still Persistence? Electoral Change in Western Europe, 1948-1979', in H. Daalder and P. Mair (eds), *Western European Party Systems*, Sage, London, pp. 67-94

Mahon, D. and Constable, M. (1973) *No Place in the Country: A Report on Second Homes in England and Wales*, Shelter, London

Maillat, D. (1978) 'Economic Growth', in Council of Europe, *Population Decline in Europe*, Arnold, London, pp. 72-92

— (1982) *Technology: A Key Factor for Regional Development*, Georgi, Geneva

Mann, P.H. (1965) *An Approach to Urban Sociology*, Routledge & Kegan Paul, London

Maos, J.O. (1981) 'Land Settlement in Mediterranean Region', *Ekistics, 290*, 382-93

Marcellesi, J-B. (1979) 'Quelques problèmes de l'hégémonie culturelle en France: langue nationale et langues régionales', *International Journal of the Sociology of Language, 21*, 63-80

Marchant, E.C. (1970) *The Countries of Europe as Seen by Their Geographers*, Harrap, London

Marquand, J. (1980) 'Spatial Change and Economic Divergence in the EEC', *Journal of Common Market Studies, 19*, 1-20

Martin, T. (1981) *Some Aspects of the Political Geography of Community Councils in Scotland*, Research Seminar Paper no. 11, Dept. of Geography, University of Strathclyde, Glasgow

Martinotti, G. (1981) 'The Illusive Autonomy: Central Control and Decentralisation in the Italian Local Financial System', in L.J. Sharpe (ed.), *The Local Fiscal Crisis in Western Europe*, Sage, Beverly Hills, pp. 63-124

Martins, M.R. and Mawson, J. (1980) 'The Evolution of EEC Regional Policy: Cosmetics or Major Surgery', *Local Government Studies, 6*, 29-56

— and — (1981) 'The Revision of the European Regional Development Fund Regulation', *Built Environment, 7*, 190-9

— and — (1982) 'The Programming of Regional Development in the European Community', *Journal of Common Market Studies, 20*, 229-44

Masser, F.I. and Stroud, D.C. (1965) 'The Metropolitan Village', *Town Planning Review, 36*, 111-24

Massey, D. and Catalano, A. (1978) *Capital and Land*, Arnold, London

— and Meegan, R. (1982) *The Anatomy of Job Loss*, Methuen, London

Maunder, P. (ed.) (1979) *Government Intervention in the Developed Economy*, Croom Helm, London

Mauret, E. (1974) *Aménagement, urbanisme, paysage*, Dunod, Paris

Mayer, K. (1975) 'Intra-European Migration During the Past 20 Years', *International Migration Review, 9*, 441-7

Mayer, W. (1978) 'City Construction in Vienna at the Turn of the Century', *Jarbuch des Vereins für Geschichte der Stadt Wien, 34*, 276-308

Mead, W.R. (1982) 'The Discovery of Europe', *Geography, 67*, 193-202

Medhurst, K. (1981) 'Basques and Basque Nationalism', in C.H. Williams (ed.), *National Separatism*, University of Wales Press, Cardiff

Meitzen, A. (1895) *Siedlung und Agrarwesen der Westgermanen und Ostgermanen, der Keten, Römer, Finnen und Slawen*, Wilhelm Hertz, Munich

Mellor, R.E.H. and Smith, E.A. (1979) *Europe: A Geographical Survey of the Continent*, Macmillan, London

Mendras, H. (1970) *The Vanishing Peasant*, MIT Press, Cambridge, Mass.

— (1979) *Voyage au pays de l'utopie rustique*, Actes/Sud, Paris

Mensch, G. (1983) *Stalemate in Technology: Innovations Overcome the Depression*, Ballinger, Cambridge, Mass.

Merritt, R.L. (1968) 'Visual Representation of Mutual Friendliness', in R. Merritt and D. Puchala (eds), *Western European Perspectives on International Affairs: Public Opinion Studies and Evaluations*, Praeger, New York, pp. 111-41

Metcalf, D. (1969) *The Economics of Agriculture*, Penguin, Harmondsworth

Mik, G. (1983) 'Residential Segregation in Rotterdam: Background and Policy', *Tijdschrift voor Economische en Sociale Geografie, 74*, 74-86

Miliband, R. (1973) *The State in Capitalist Society*, Quartet, London

Millar, A.R. (1980) *A Study of Multiply Derived Households in Scotland*, Central

Research Unit, Scottish Office, Edinburgh

Millas, A. (1980) 'Planning for the Elderly Within the Context of a Neighbourhood', *Ekistics*, *47*, 273-6

Miller, M. (1982) 'The Political Impact of Foreign Labour: a Re-evaluation of Western European Experience', *International Migration Review*, *16*, 27-60

Mintz, S. (1973) 'A Note on the Definition of Peasantries', *Journal of Peasant Studies*, *1*, 91-106

Molle, W. and Paelinck, J. (1979) 'Regional Policy', in P. Coffey (ed.), *Economic Policies of the Common Market*, Macmillan, London, pp. 146-77

—–, Holst, B. van and Smit, H. (1980) *Regional Disparity and Economic Development in the European Community*, Saxon House, Farnborough

Monnier, A. (1981) 'L'Europe et les pays développés d'outre-mer: données statistiques', *Population*, *36*, 885-96

Moore, B. and Rhodes, J. (1975) 'The Economic and Exchequer Implications of British Regional Policy', in J. Vaizey (ed.), *Economic Sovereignty and Regional Policy*, Gill and Macmillan, Dublin, pp. 80-102

— and — (1982) 'A Second Great Depression in the UK: Can Anything Be Done?', *Regional Studies*, *16*, 323-33

Moore, D. (ed.) (1979) *Disadvantaged Rural Europe*, Arkleton Trust, Langholm

Moore, R. (1980) 'Industrial Growth and Development Policies in the British Periphery', in F. Buttell and A. Newby (eds), *Rural Sociology of the Advanced Societies*, Croom Helm, London, 387-404

Moreno, M. and Miguel, A. de (1978) *La estructura social de las ciudades españolas*, Centro de Investigaciones Sociologicas, Madrid

Morgado, N.A. (1979) 'Portugal', in W.R. Lee (ed.), *European Demography and Economic Growth*, Croom Helm, London, pp. 319-39

Mormont, M. (1983) 'The Emergence of Rural Struggles and Their Regional Effect', *International Journal of Urban and Regional Research*, *7*, 559-76

Morris, N. (1980) The Common Agricultural Policy, *Fiscal Studies*, *1*, 17-35

Morrison, R. (1977) *The Perception of Poverty in Europe*, Commission of the European Communities, Brussels

Moseley, M.J. (1974) *Growth Centres in Spatial Planning*, Pergamon, Oxford

— (ed.) (1978) *Social Issues in Rural Norfolk*, Centre for East Anglian Studies, Norwich

— (1979) *Accessibility: The Rural Challenge*, Methuen, London

— (1980) 'Is Rural Deprivation Really Rural?', *The Planner*, *66*, 97

Mottura, G. and Pugliese, E. (1980) 'Capitalism in Agriculture and Capitalistic Agriculture: The Italian Case', in F. Buttel, and H. Newby (eds), *The Rural Sociology of Advanced Societies*, Croom Helm, London, pp. 171-200

Mughan, A. (1979) 'Modernisation and Regional Relative Deprivation: Towards a Theory of Ethnic Conflict', in L.J. Sharpe (ed.), *Decentralist Trends in Western Democracies*, Sage, London, pp. 279-312

Muir, R. (1983) 'Regional Administration in Western Europe: an Interpretation of Factors Leading to Decentralisation of Political Power', in M.A. Busteed (ed.), *Developments in Political Geography*, Academic Press, London

Müller, J.O. (1964) *Die Einstellung zur Landarbeit in bäuerlichen Familienbetrieben*, Forschungsgesellschaft für Agrarpolitik und Agrarsociologie E.V., Bonn

Mullin, J.R. (1981) 'The Architecture of the Third Reich', *Journal of the American Planning Association*, *47*, 35-47

— (1982) 'Ideology, Planning Theory and the German City in the Inter-war Years', *Town Planning Review*, *53*, 115-30, 257-72

Mumford, L. (1938) *The Culture of Cities*, Secker & Warburg, London

— (1966) *The City in History*, Penguin, Harmondsworth

Muñoz-Perez, F. (1982) 'L'évolution de la fecondité dans les pays industrialisés depuis 1971', *Population*, *37*, 483-512

Münzer, L. (1983) 'Recent Changes in Agricultural Structure and Rural Settlement in Peripheral Mountain Areas of Hedmark County, eastern Norway', *Norsk Geografisk Tijdsskrift*, *37*, 81-110

Myklebost, H. (1981) 'Regional Variations in Mortality in Norway', *Fennia*, *159*, 153-63

Myrdal, G. (1957) *Economic Theory and Underdeveloped Regions*, Duckworth, London

Nairn, T. (1977) *The Break-Up of Britain*, New Left Books, London

Nash, R. (1980) *Schooling in Rural Societies*, Methuen, London

Navarro, V. (1982) 'The Crisis of the International Capitalist Order and its Implications for the Welfare State', *International Journal of Health Services*, *12*, 169-90

Naylon, J. (1981) 'Barcelona', in M. Pacione (ed.), *Urban Problems and Planning in the Developed World*, Croom Helm, London, pp. 223-57

—— (1982) *Politics and Urban Growth in Franco Spain: the Case of Barcelona*, University of Keele, Department of Geography, Occasional Paper no. 2

Neate, S. (1981) *Rural Deprivation*, Bibliography no. 8, Geo. Abstracts, Norwich

Nellner, W. (1976) 'Die Innere Gliederung Stadtischer Siedlungs-agglomerationen', *Forschungs und Sitzungsberichte Akademie fur Raumforschung and Landesplanung*, *112*, 35-74

Netting, R. McC. (1981) *Balancing on an Alp*, Cambridge University Press, Cambridge

Newby, H. (1978) 'The Rural Sociology of Advanced Capitalist Societies', in H. Newby (ed.), *International Perspectives in Rural Sociology*, Wiley, London, pp. 3-30

—— (1979) *Green and Pleasant Land?*, Penguin, Harmondsworth

—— (1980) 'Rural Sociology', *Current Sociology*, *28*, 1-141

——, Bell, C., Rose, D. and Saunders, P. (1978) *Property, Paternalism and Power*, Hutchinson, London

Newton, K. (1976) 'Community Performance in Britain', in M.S. Archer (ed.), *Problems of Current Sociological Research*, Current Sociology, *22*, Mouton, The Hague, pp. 49-86

—— (1984) 'Public Services in Cities and Counties', in A. Kirby, P.L. Knox and S. Pinch (eds), *Public Service Provision and Urban Development*, Croom Helm, London, pp. 19-43

—— *et al.* (1980) *Balancing the Books*, Sage, Beverly Hills

Nicholson, B. (1976) 'Return Migration to a Rural Area: an Example from Norway', *Sociologia Ruralis*, *14*, 227-44

Nicol, W.R. (1982) 'Estimating the Effects of Regional Policy: A Critique of the European Experience', *Regional Studies*, *16*, 199-210

—— and Yuill, D. (1980) *Regional Problems and Policies in Europe: The Postwar Experience*, Studies in Public Policy no. 53, Centre for the Study of Public Policy, University of Strathclyde, Glasgow

Nikolinakos, M. (1973) *Politische Ökonomie der Gastarbeiterfrage: Migration und Kapitalismus*, Rowohlt, Hamburg

Nordlinger, E.A. (1967) *The Working Class Tories*, MacGibbon & Kee, London

Nutter, G.W. (1978) *Growth of Government in the West*, American Enterprise Institute, Washington, DC

OECD (1975) 'A Turning Point for European Migration', *OECD Observer*, *76*, 13-16

OECD (1976a) *Regional Problems and Policies in OECD Countries*, vol. 1, OECD, Paris

OECD (1976b) *Regional Problems and Policies in OECD Countries*, vol. 2, OECD,

Paris

OECD (1976c) *Public Expenditure on Income Maintenance Programmes*, OECD, Paris

OECD (1980) 'Social Change in OECD Countries 1950-1980', *OECD Observer*, *107*, 21-35

OECD (1981a) *The Welfare State in Crisis: An Account of the Conference on Social Policies in the 1980s*, OECD, Paris

OECD (1981b) *Economic Survey: Portugal*, OECD, Paris

O'Farrell, P.N. (1980) 'Multinational Enterprise and Regional Development – Irish Evidence', *Regional Studies*, *12*, 131-50

O'Flanagan, T.P. (1980) 'Agrarian Structures in Northwest Iberia', *Geoforum*, *11*, 157-69

Ogden, P.E. and Huss, M-N. (1982) 'Demography and Pro-natalism in France in the Nineteenth and Twentieth Centuries', *Journal of Historical Geography*, *8*, 283-98

Ohlin, G. (1967) 'Regulation démographique et développement économique', OECD, Paris

O'Loughlin, J. (1980a) 'District Size and Party Strength: a Comparison of Sixteen Democracies', *Environment and Planning*, *12A*, 247-62

—— (1980b) 'Distribution and Migration of Foreigners in German Cities', *Geographical Review*, *70*, 253-76

—— (1983) 'Spatial Inequalities in Western Cities: a Comparison of North American and German Urban Areas', *Social Indicators Research*, *13*, 185-212

Orridge, A. and Williams, C.H. (1982) 'Autonomist Nationalism: A Theoretical Framework for Spatial Variations in its Genesis and Development', *Political Geography Quarterly*, *1*, 19-39

Pacione, M. (1982a) 'The Viability of Smaller Rural Settlements', *Tijdschrift voor Economische en Sociale Geografie*, *73*, 149-61

—— (1982b) 'Space Preferences, Locational Decisions and the Dispersal of Civil Servants from London', *Environment and Planning*, *14A*, 323-33

Paddison, R. (1983) *The Fragmented State*, Blackwell, London

Pahl, R.E. (1965) *Urbs in Rure*, Weidenfeld & Nicolson, London

Paine, S. (1974) *Exporting Workers: The Turkish Case*, Cambridge University Press, London

—— (1979) 'Replacement of the West European Migrant Labour System by Investment in the European Periphery', in D. Seers, B. Schaffer and M-L. Kiljunen (eds), *Underdeveloped Europe*, Harvester Press, Hassocks, Sussex, pp. 65-96

Parker, G. (1979) *The Countries of Community Europe: A Geographical Survey of Contemporary Issues*, Macmillan, London

Parkin, F. (1967) 'Working-class Conservatives: A Theory of Political Deviance', *British Journal of Sociology*, *18*, 278-90

Payne, S. (1971) 'Catalan and Basque nationalism', *Journal of Contemporary History*, *6*, 15-51

Peach, C. (1975) 'Immigrants in the Inner City', *Geographical Journal*, *141*, 372-9

Peacock, A.T. and Wiseman, J. (1967) *The Growth of Public Expenditure in the UK*, Allen & Unwin, London

Pedersen, M.N. (1983) 'Changing Patterns of Electoral Volatility in European Party Systems 1948-1977: Explorations in Explanation', in H. Daalder and P. Mair (eds), *Western European Party Systems*, Sage, London, pp. 29-66

Pen, J. (1979) 'A Clear Case of Levelling: Income Equalisation in the Netherlands', *Social Research*, *46*, 682-94

Penouil, M. and Petrella, R. (1982) *The Location of Growing Industries in Europe*,

European Coordination Centre for Research and Documentation in Social Sciences, Vienna

Pérez-Díaz, V.M. (1976) 'Process of Change in Rural Castilian Communities', in J.B. Aceves and W.A. Douglass (eds), *The Changing Faces of Rural Spain*, Shenkman, Cambridge, Mass., pp. 123-42

Perrons, D. (1981) 'The Role of Ireland in the New International Division of Labour', *Regional Studies*, *15*, 81-100

Peters, B.G. (1979) 'Dimensions of Quality of Life in an Urban Area: an Analysis of Stockholm', in M.C. Romanos (ed.), *Western European Cities in Crisis*, Lexington, Lexington, Mass., pp. 31-46

Philip, A.B. (1980) 'European Nationalism in the Nineteenth and Twentieth Centuries', in R. Mitchinson (ed.), *The Roots of Nationalism: Studies in Northern Europe*, John Donald, Edinburgh, pp. 1-10

—— (1982) 'Pressure Groups and Policy Formation in the European Communities', *Policy and Politics*, *10*, 459-76

Pierce, S. (1980) 'Single Mothers and the Concept of Female Dependency in the Development of the Welfare State in Britain', *Journal of Comparative Family Studies*, *11*, 57-86

Pierce, N.R. and Hagstrom (1981) 'Inner City in Three Countries', in G.G. Schwartz (ed.), *Advanced Industrialisation and the Inner Cities*, D.C. Heath, Lexington, Mass., pp. 141-55

Pinch, S. (1979) 'Territorial Justice in the City: a Case Study of the Social Services for the Elderly in Greater London', in D. Herbert and D. Smith (eds), *Social Problems and the City*, Oxford University Press, London, pp. 201-23

Pinder, D.A. (1978) 'Guiding Economic Development in the EEC: the Approach of the European Investment Bank', *Geography*, *63*, 88-97

—— (1983) *Regional Economic Development and Policy: Theory and Practice in the European Community*, Allen & Unwin, London

Piore, M.J. (1979) *Birds of Passage: Migrant Labour and Industrial Societies*, Cambridge University Press, Cambridge

Pirenne, H. (1925) *Medieval Cities*, Princeton University Press, Princeton

Pitfield, D.E. (1978) 'The Quest for Effective Regional Policy', *Regional Studies*, *12*, 429-43

Pizzorno, A. (1963) 'An Introduction to the Theory of Popular Participation', *Social Science Information*, *9*, 49-56

Pluta, J.E. (1978) 'National Defence and Social Welfare Budget Trends in Ten Nations of Postwar Western Europe', *International Journal of Social Economics*, *5*, 3-21

Policy Studies Institute (1979) *European Integration, Regional Devolution and National Parliaments*, European Centre for Political Studies, Policy Studies Institute, London

Pollard, S. (1981) *Peaceful Conquest*, Oxford University Press, London

Poppel, F.W.A. van (1981) 'Regional Mortality Differences in Western Europe: a Review of the Situation in the Seventies', *Social Science and Medicine*, 15D, 341-52

Poulantzas, N. (1978) *State, Power, Socialism*, New Left Books, London

Pred, A.R. (1974) *Major Job-Providing Organisations and Systems of Cities*, Association of American Geographers, Resource Paper 27, Washington, DC

Preteceille, E. (1981) 'Left-wing Local Governments and Services Policy in France', *International Journal of Urban and Regional Research*, *5*, 411-18

Price, C.A. (1963) *Southern Europeans in Australia*, Oxford University Press, Melbourne

Pringle, D. (1982) 'Regional Disparities in the Quantity of Life in the Republic of Ireland 1971-1977', *Irish Geography*, *15*, 22-34

Prud'homme, R. (1974) 'Regional Economic Policy in France', in N.M. Hansen (ed.), *Public Policy and Regional Economic Development*, Ballinger, Cambridge, Mass., pp. 33-64

Puls, W.W. (1973) 'Ballungsnahe Randzonen als Magnet: Aus dem Raumordnungsbericht 1972 der Bundesregierung', *Gegenwartskunde*, *22*, 205-12

Pumain, D. (1982) 'Chemin de fer et croissance urbaine en France au XIXe siècle', *Annales de géographie*, *91*, 529-50

—— and Saint-Julien, T. (1979) 'Recent Transformations in the French Urban System', *L'Espace géographique*, *8*, 203-10

Pyle, D. (1976) 'Aspects of Resource Allocation by Local Education Authorities', *Social and Economic Administration*, *10*, 106-21

Radford, E. (1970) *The New Villagers*, Cass, London

Ragin, C.C. (1979) 'Ethnic Political Mobilisation: the Welsh Case', *American Sociological Review*, *44*, 619-35

Ravenstein, E. (1885) 'The Laws of Migration', *Journal of the Royal Statistical Society*, *48*, 167-227

Razzell, P. (1974) 'An Interpretation of the Modern Rise of Population in Europe – a Critique', *Population Studies*, *28*, 5-17

—— (1977) *The Conquest of Smallpox: The Impact of Inoculation on Smallpox Mortality in Eighteenth Century Britain*, Caliban Books, Firle, Sussex

Reddaway, W. (1939) *The Economics of Declining Population*, Macmillan, London

Rees, A.D. (1950) *Life in a Welsh Countryside*, University of Wales Press, Cardiff

Reichenbach, H. (1980) 'A Politico-Economic Overview', in D. Seers and C. Vaitsos (eds), *Integration and Unequal Development*, Macmillan, London, pp. 75-99

Rex, J. (1982) 'The 1981 Urban Riots in Britain', *International Journal of Urban and Regional Research*, *6*, 99-114

Rhoades, R. (1978) 'Intra-European Return Migration and Rural Development: Lessons from the Spanish Case', *Human Organization*, *37*, 136-47

—— (1980) 'European Cyclical Migration and Economic Development: the Case of Southern Spain', in G. Gmelch and W.P. Zenner (eds), *Urban Life*, St. Martin's, New York, pp. 106-32

Richardson, H.W. (1969) *Elements of Regional Economics*, Penguin, Harmondsworth

—— (1975) *Regional Development Policy and Planning in Spain*, Saxon House, Farnborough

Richmond, H.A. (1968) 'Return Migration from Canada to Britain', *Population Studies*, *22*, 263-71

Ringen, S. (1974) 'Welfare Studies in Scandinavia', *Scandinavian Political Studies*, *9*, 187-96

Riquet, P. (1978) 'Clivages sociaux au sein des espaces urbanisés en Allemagne Fédérale', *Bulletin, Assoc. géographique Française*, *449*, 13-21

Rist, R.C. (1979) 'Migration and Marginality: Guestworkers in Germany and France', *Daedalus*, *108*, 95-108

Robson, B. (1975) *Urban Social Areas*, Oxford University Press, London

Rodgers, A.W. (1983) 'Rural Housing', in M. Pacione (ed.), *Progress in Rural Geography*, Croom Helm, London

Rogers, S.T. and Davey, B.H. (eds) (1973) *The Common Agricultural Policy and Britain*, Saxon House, Farnborough

Rokkan, S. (1967) 'Geography, Religion and Social Class: Crosscutting Cleavages in Norwegian Politics', in S.M. Lipset and S. Rokkan (eds), *Party Systems and Voter Alignments: Cross-National Perspectives*, Collier-Macmillan, London, pp. 367-444

Rokkan, S. (1970) *Citizens, Elections, Parties,* McKay, New York
— (1980) 'Territories, Centres and Peripheries', in J. Gottmann (ed.), *Centre and Periphery: Spatial Variation in Politics,* Sage, Beverly Hills, pp. 163-204
— and Urwin, D. (1982) 'Centres and Peripheries in Western Europe', in S. Rokkan and D. Urwin (eds), *The Politics of Territorial Identity,* Sage, London, pp. 1-18
Romanos, M.C. (1979) 'Forsaken Farms: the Village-to-city Movement in Western Europe', in M.C. Romanos (ed.), *Western European Cities in Crisis,* Lexington Books, Lexington, Mass., pp. 3-19
Room, G. (1982) 'Understanding Poverty', in J. Dennett *et al., Europe Against Poverty,* Bedford Square Press, London, pp. 163-84
Rose, A.M. (1969) *Migrants in Europe: Problems of Acceptance and Adjustment,* University of Minnesota Press, Minneapolis
Rose, R. and Urwin, D.W. (1970) 'Persistence and Change in Western Party Systems Since 1945', *Political Studies, 18,* 287-319
— and — (1975) *Regional Differentation and Political Unity in Western Nations,* Sage Professional Papers, Contemporary Political Sociology Series, *06-007,* Sage Publications, Beverly Hills
Rothman, S., Scarrow, H. and Schain, M. (1976) *European Society and Politics,* West Publishing, St Paul, Minn.
Rothwell, R. (1982) 'The Role of Technology in Industrial Change: Implications for Regional Policy', *Regional Studies, 16,* 361-7
Roussel, L. and Festy, P. (1979) *Recent Trends in Attitudes and Behaviour Affecting the Family in Council of Europe Member States,* Population Studies, *4,* Council of Europe, Strasbourg
Rowley, G. (1975) 'Electoral Change and Reapportionment: Prescriptive Ideals and Reality', *Tijdschrift voor Economische en Sociale Geografie, 66,* 108-20
Rugg, D.S. (1979) *Spatial Foundations of Urbansim* (2nd edition), W.C. Brown, Dubuque, Iowa
Ryder, N.B. (1975) 'Notes on Stationary Populations', *Population Index, 41,* 3-28
Sacchi de Angelis, M.E. (1979) 'Rural Tourism in Umbria', in C. Christians and J. Claude (eds), *Recherches de géographie rurale,* Bulletin de la Société Géographique de Liège, Liège, pp. 809-24
Sallnow, J. and John, A. (1982) *An Electoral Atlas of Europe 1968-1981,* Butterworth Scientific, London
Saloutos, T. (1956) *They Remember America: the Study of Repatriated Greek-Americans,* University of California Press, Berkeley
Salt, J. (1976) 'International Labour Migration: the Geographical Pattern of Demand', in J. Salt and H. Clout (eds), *Migration in Post-War Europe,* Oxford University Press, London, pp. 80-125
— and Clout, H. (1976) 'International Labour Migration: the Sources of Supply', in J. Salt and H. Clout (eds), *Migration in Post-War Europe,* Oxford University Press, London
Sandre, P. de (1978) 'The Influence of Governments', in Council of Europe, *Population Decline in Europe,* Arnold, London
Scase, R. (1980) 'Introduction', in R. Scase (ed.), *The State in Western Europe,* Croom Helm, London, pp. 11-22
Schäfer, G. (1981) 'Trends in Local Government Finance in the Federal Republic of Germany', in L.J. Sharpe (ed.), *The Local Fiscal Crisis in Western Europe,* Sage, Beverly Hills, pp. 229-69
Schiller, G. (1974) 'Auswirkungen der Arbeitskräftewanderung in den Herkunftsländern', in R. Lohrmann and K. Manfass (eds), *Auslanderbeschäftigung und Internationale Politik,* Oldenburger Verlag, Munich

Schmidt, M.G. (1983) 'The Welfare State and the Economy in Periods of Economic Crisis: A Comparative Study of 23 OECD Nations', *European Journal of Political Research, 11,* 1-26

Schmitthoff, C.M. (1973) 'The Multinational Enterprise in the UK', in J.G. Smith and R.W. Wright (eds), *Nationalism and the Multinational Enterprise,* Sijthoff, The Hague, pp. 28-47

Schofield, J.A. (1976) 'Economic Efficiency and Regional Policy', *Urban Studies, 13,* 181-92

Schumpeter, J.A. (1939) *Business Cycles,* McGraw-Hill, New York

Secchi, C. (1982) 'Impact on the Less Developed Regions of the EEC', in D. Seers and C. Vaitsos (eds), *The Second Enlargement of the EEC,* Macmillan, London, pp. 176-89

Seers, D. (1979) 'The Periphery of Europe', in D. Seers, B. Schaffer and M.-L. Kiljunen (eds), *Underdeveloped Europe: Studies in Core-Periphery Relations,* Harvester Press, Hassocks, Sussex, pp. 3-34

Seton-Watson, H. (1977) *Nations and States: An Enquiry into the Origins of Nations and the Politics of Nationalism,* Methuen, London

Sewell, W.H. (1950) 'Needed Research in Rural Sociology', *Rural Sociology, 15,* 115-30

Shanin, T. (1971) *Peasants and Peasant Societies: Selected Readings,* Penguin, Harmondsworth

Sharpe, L.J. (1979) 'Decentralist Trends in Western Democracies: A First Appraisal', in L.J. Sharpe (ed.), *Decentralist Trends in Western Democracies,* Sage, Beverly Hills, pp. 9-80

— (1980) 'Is There a Fiscal Crisis in Western European Local Government?', *International Political Science Review, 1,* 203-26

Shaw, M. (ed.) (1979) *Rural Deprivation,* Geo Books, Norwich

Sheahan, J.B. (1976) 'Experience with Public Enterprise in France and Italy', in W.G. Shepherd (ed), *Public Enterprise: Economic Analysis of Theory and Practice,* D.C. Heath, Lexington, Mass., pp. 123-83

Sher, J.P. (1978) 'Education in Sparsely-populated Areas of Developed Nations', *Educational Forum, 63,* 83-8

Shils, E. (1981) 'Center and Periphery', in *Essays in Memory of M. Polyani,* Routledge & Kegan Paul, London, pp. 117-30

Shoard, M. (1980) *The Theft of the Countryside,* Temple Smith, London

Short, J. (1981) *Public Expenditure and Taxation in the UK,* Gower, Farnborough

— (1982), 'Urban Policy and British Cities', *Journal of the American Planning Association, 48,* 39-52

Signorelli, A. (1980) 'Regional Policies in Italy for Migrant Workers Returning Home', in R.D. Grillo (ed.), *'Nation' and 'State' in Europe: Anthropological Perspectives,* Academic Press, London, pp. 89-104

Smailes, A.E. (1966) *The Geography of Towns,* Hutchinson, London

Smidt, M. de (1979) 'Regional Policy and Planning in the Netherlands', *L'Espace géographique, 8,* 161-70

Smith, D.M. (1979) *Where the Grass is Greener,* Penguin, Harmondsworth

Smith, D.S. (1977) 'A Homeostatic Demographic Regime: Patterns in West European Family Reconstitution Studies', in R.D. Lee (ed.), *Population Patterns in the Past,* Academic Press, New York, pp. 19-51

Smith, G. (1976) 'Social Movements and Party Systems in Western Europe', in M. Kolinsky and W.E. Paterson (eds), *Social and Political Movements in Western Europe,* Croom Helm, London, pp. 331-54

— (1980) *Politics in Western Europe: A Comparative Analysis,* Heinemann, London

Smith, I.J. (1979) 'The Effect of External Takeovers on Manufacturing Employment Change in the Northern Region between 1963 and 1973', *Regional Studies, 13,* 421-37

Solomos, J. (ed.) (1982) *Migrant Workers in Metropolitan Cities,* European Science Foundation, Strasbourg

Sommers, L.M. (1983) 'Cities of Western Europe', in S.D. Brunn and J.F. Williams (eds), *Cities of the World,* Harper & Row, New York, pp. 84-121

SOU (1975) *Kommunal Demokrati,* Statens Offentliga Utredningar, Stockholm

Stammen, T. (1980) *Political Parties in Europe,* John Martin Publishing, London

Standing Conference of Rural Community Council (1978) *The Decline of Rural Services,* National Council of Social Services, London

Stanyer, J. (1976) *Understanding Local Government,* Fontana, London

Stearns, P.N. (1977) *Old Age in European Society,* Croom Helm, London

Stephens, M. (1976) *Linguistic Minorities in Western Europe,* Gomer Press, Llandysal

Stephenson, G. (1972) 'Cultural Regionalism and the Unitary State Idea in Belgium', *Geographical Review, 62,* 501-23

Strayer, J.R. (1963) 'The Historical Experience of Nation-building in Europe', in K.W. Deutsch and W.J. Folz, (eds), *Nation Building,* Atherton Press, New York, pp. 139-57

Streit, M.E. (1977) 'Government and Business: the Case of West Germany', in R.T. Griffiths (ed.), *Government Business and Labour in European Capitalism,* Europotentials Press, London, pp. 120-34

Sundberg, N. and Öström, K. (1982) 'Migration and Welfare: a Study in North Sweden', *Geografiska Annaler, 64B,* pp. 153-60

Swann, D. (1983) *Competition and Industrial Policy in the European Community,* Methuen, London

Swanson, J. (1979) 'The Consequences of Emigration for Economic Development: a Review of the Literature', *Papers in Anthropology, 20,* 39-56

Tarn, J.N. (1973) *Five Per Cent Philanthropy,* Cambridge University Press, Cambridge

Tarrow, S. (1978) 'Introduction', in S. Tarrow, P.J. Katzenstein and L. Graziano (eds), *Territorial Politics in Industrial Nations,* Praeger, New York, pp. 1-27

Tarschys, D. (1975) 'The Growth of Public Expenditures; Nine Modes of Explanation', *Scandinavian Political Studies, 10,* 1-28

Taylor, P. (1982) *The Limits of European Integration,* Croom Helm, London

Taylor, P.J. and Johnston, R.J. (1979) *Geography of Elections,* Penguin, Harmondsworth

Taylor, R. (1979) 'Migration and the Residual Community', *Sociological Review, 27,* 475-89

Terán, F. de (1981) 'New Planning Experiences in Democratic Spain', *International Journal of Urban and Regional Research, 5,* 96-105

Ter Hoeven, P.J.A. (1978) 'The Social Bases of Flemish Nationalism', *International Journal of the Sociology of Language, 15,* 21-32

Thieme, G. (1983) 'Agricultural Change and its impact in Rural Areas', in T. Wild (ed.), *Urban and Rural Change in West Germany,* Croom Helm, London, pp. 220-47

Thomas, S. and Tuppen, J. (1977) 'Readjustment in the Ruhr – the Case of Bochum', *Geography, 62,* 168-75

Thomson Report (1973) *Report on the Regional Problem in the Enlarged Community,* Commission of the European Communities, Brussels

Thorns, D.C. (1968) 'The Changing System of Rural Stratification', *Sociological Review, 8,* 161-78

Thumerelle, P.J. (1981) *Migrations internes et externes en Europe occidentale*, Hommes et Terres du Nord, Villeneuve d'Ascq. (2 vols.)

Tilly, C. (ed.) (1975) *The Formation of National States in Western Europe*, Princeton University Press, Princeton

Tivey, L. (ed.) (1981) *The Nation State*, Robertson, London

Townsend, A.R. (1982) *The Impact of Recession*, Croom Helm, London

Townsend, P. (1979) *Poverty in the United Kingdom*, Penguin, Harmondsworth

Tracy, M. (1982) *Agriculture in Western Europe: Challenge and Reponse 1880-1980*, Granada, St Albans

Trébous, M. (1980) *Migration and Development: the Case of Algeria*, OECD, Paris

Treves, A. (1981) 'La Politique anti-urbaine fasciste et un siècle de résistance contre l'urbanisation en Italie', *L'Espace géographique*, *10*, 115-24

Trevor-Roper, H.R. (1968) 'The Phenomenon of Fascism', in S.J. Woolf (ed.), *European Fascism*, Weidenfeld & Nicolson, London

Triffin, R. (1980) 'The European Monetary System', in D. Seers and C. Vaitsos (eds), *Integration and Unequal Development*, Macmillan, London, pp. 223-33

Trilling, J. (1981) 'French Environmental Politics', *International Journal of Urban and Regional Research*, *5*, 67-82

Tschudi, A.B. (1979) 'Multipurpose Planning of Recreational Areas Surrounding Oslo', in G. Christians and J. Claude (eds), *Recherches de géographie rurale*, Bulletin de la Société Géographique de Liège, Liège, pp. 839-53

Tuppen, J.N. (1980) *France*, Dawson, Folkestone

— (1983) 'The development of French New Towns: an Assessment of Progress', *Urban Studies*, *20*, 11-30

Tykkäläinen, M. (1981) 'Accessibility in the Provinces of Finland', *Fennia*, *159*, 361-96

Unger, K. (1981) 'Greek Emigration to and Return from West Germany', *Ekistics*, *48*, 369-74

United Kingdom Select Committee on the European Communities (1980) *Policies for Rural Areas in the European Community*, 27th Report, Session 1979/80, HMSO, London

United Nations (1975) *Economic Survey of Europe in 1974, Part II: Post-war demographic trends in Europe and the outlook until the year 2000*, United Nations, New York

— (1980) *Compendium of Social Statistics (1977)*, Statistical Papers, Series K, No. 4, United Nations Organisation, New York

Urry, J. (1984) 'Capitalist Restructuring, Recomposition and the Regions', in A. Bradley and P. Lowe (eds), *Locality and Rurality: Economy and Society in Rural Regions*, Geo Books, Norwich

Urwin, D. (1983) 'Harbinger, Fossil or Fleabite? "Regionalism" and the West European Party Mosaic', in H. Daalder and P. Mair (eds), *Western European Party Systems*, Sage, London

Vagts, S. (1960) 'Deutsch-Amerikanische Rückwanderung', *Beihefte zum Jahrbuch für Amerikastudien*, *No. 6*, Heidelberg

Vaitsos, C. (1979) 'Transnational Corporations and Europe', in D. Seers, B. Schaffer and M.-L. Kiljunen (eds), *Underdeveloped Europe*, Harvester Press, Hassocks, Sussex, pp. 97-102

Vallaux, C. (1908) *Géographie sociale*, Mouton, Paris

Vanderkamp, J. (1970) 'The Effects of Outmigration on Regional Employment', *Canadian Journal of Economics*, *3*, 541-50

Vane, R. de (1975) *Second Home Ownership: A Case Study*, Bangor Occasional Papers in Economics No. 6, University of Wales, Cardiff

Vanhove, N. and Klaasen, L.H. (1980) *Regional Policy: A European Approach*, Saxon House, Farnborough

Vaughan, R. (1979) *Twentieth Century Europe: Paths to Unity*, Croom Helm, London

Veldman, J. (1981) 'Space and Society in Rural Areas', paper presented to the 11th European Congress for Rural Sociology, Helsinki

Verwayen, H. (1980) 'The Specification and Measurement of the Quality of Life in OECD countries', in A. Szalai and F.M. Andrews (eds), *The Quality of Life: Comparative Studies*, Sage Studies in International Sociology, *20*, Sage, Beverly Hills, pp. 235-47

Vincent, J. (1978) *The Political Economy of Alpine Development*, Exeter Research Group, Discussion Paper No. 3, University of Exeter

Vining, D.R., Jr. and Kontuly, T. (1978) 'Population Dispersal from Major Metropolitan Regions: an International Comparison', *International Regional Science Review, 3*, 49-73

— (1982) 'Migration Between the Core and the Periphery', *Scientific American, 247*, 44-53

Vogel, J. (1981) *Social Report on Inequality in Sweden: Distribution of Welfare at the End of the 1970s*, National Central Bureau of Statistics, Stockholm

Vries, J. de (1976) *Economy of Europe in an Age of Crisis, 1600-1750*, Cambridge University Press, Cambridge

Wade, R. (1980) 'The Italian State and the Underdevelopment of South Italy', in R.D. Grillo (ed.), *'Nation' and 'State' in Europe*, Academic Press, London, pp. 151-72

Wagtskjold, J.R. (ed.) (1982) *Komparative analyser au regionale data for de nordiskeland*, Universitetsforlaget, Oslo

Wainwright, H. (1978) 'Women and the Division of Labour', in P. Abrams (ed.), *Work, Urbanism and Inequality*, Weidenfeld & Nicolson, London

Wallace, D.B. (1981) 'Rural Policy: A Review Article', *Town Planning Review, 52*, 215-22

Walle, E. de and Knodel, J. (1967) 'Demographic Transition and Fertility Decline: the European Case', in International Union for the Scientific Study of Population, *Contributed Papers to 1967 Sydney Conference*, pp. 47-55

— and — (1980) 'Europe's Fertility Transition: New Evidence and Lessons for Today's Developing World', *Population Bulletin, 34*, 6-18

Wallerstein, E. (1974) *The Modern World System*, Academic Press, London

Walter-Busch, E. (1983) 'Subjective and Objective Indicators of Regional Quality of Life in Switzerland', *Social Indicators Research, 12*, 337-92

Watkins, R. (1979) 'Educational Disadvantage in Rural Areas', in M. Shaw (ed.), *Rural Deprivation and Planning*, Geo Books, Norwich, pp. 93-101

Watkins, S.C. (1981) 'Regional Patterns of Nuptiality in Europe 1870-1960', *Population Studies, 35*, 199-215

Welsenes, C. van (1979) 'The Influence of Modernisation of Agriculture on Patterns of Culture and Landscape', in C. Christians and J. Claude (eds), *Recherches de géographie rurale*, Bulletin de la Société Géographique de Liège, Liège, 473-99

Watts, D. (1981) *The Branch Plant Economy*, Longman, London

Webb, K. (1978) *The Growth of Nationalism in Scotland*, Penguin, Harmondsworth

Westoff, C.F. (1983) 'Fertility Decline in the West: Causes and Prospects', *Population and Development Review, 9*, 99-104

White, P.E. (1976) 'Tourism and Economic Development in the Rural Environment', in R. Lee and P.E. Ogden (eds), *Economy and Society in the EEC*, Saxon House, Farnborough, pp. 150-60

White, P.E. (1980) 'Migration Loss and the Residual Community', in P.E. White and R. Woods (eds), *The Geographical Impact of Migration*, Longman, London, pp. 198-222

Wiest, R.E. (1979) 'Anthropological Perspectives on Return Migration: a Critical Commentary', *Papers in Anthropology, 20*, 167-86

Wild, T. (1983a) 'Social Fallow and its Impact on the Rural Landscape', in T. Wild (ed.), *Urban and Rural Change in West Germany*, Croom Helm, London, 200-19

— (1983b) 'The Residential Dimension to Rural Change' in T. Wild (ed.), *Urban and Rural Change in West Germany*, Croom Helm, London, pp. 161-99

Williams, A.M. (1981) 'Bairros Clandestinos: Illegal Housing in Portugal', *Geografisch Tijdscrift, 15*, 24-34

Williams, C.H. (1980) 'Ethnic Separatism in Western Europe', *Tijdschrift voor Economische en Sociale Geografie, 71*, 142-58

Williams, G. (1980) 'Wales: Cultural Bases of Nineteenth and Twentieth Century Nationalism', in R. Mitchinson (ed.), *The Roots of Nationalism: Studies in Northern Europe*, John Donald, Edinburgh, pp. 119-30

Williams, P.M. (1970) *French Politics and Elections, 1931-1969*, Cambridge University Press, Cambridge

Williams, R.L. (1973) *The Country and the City*, Chatto & Windus, London

Williams, R. (1979) 'The Multinational Enterprise: A 1977 Perspective', in J. Hayward and R.N. Berki (eds), *State and Society in Contemporary Europe*, Martin Robertson, London, pp. 237-52

Williams, W.M. (1956), *The Sociology of an English Village: Gosforth*, Routledge & Kegan Paul, London

Winter, J.M. (1980) 'The Fear of Population Decline in Western Europe, 1879-1940', in R.W. Hiorns (ed.), *Demographic Patterns in Developed Societies*, Taylor & Francis, London, pp. 173-97

Wirz, H.M. (1977) 'Economics of Welfare: The Implication of Demographic Change for Europe', *Futures, 9*, 45-52

Wise, M.J. and Thorpe, P. (1970) 'The Growth of Birmingham 1800-1950', in M.J. Wise (ed.), *Birmingham and Its Regional Setting: A Scientific Survey*, County Historical Reprints, Birmingham, pp. 213-28

Wise, M. (1983) *The Common Fisheries Policy of the European Community*, Methuen, London

Wolfe, T. (1981) *From Bauhaus to Our House*, Farrar Straus Giroux, New York

Woodruffe, B.J. (1976) *Rural Settlement Policies and Plans*, Oxford University Press, Oxford

Wrochno-Stanke, K. (1982) 'Immigrant Workers in Western Europe and European Integration', in B. De Marchi and A.M. Boileau (eds), *Boundaries and Minorities in Western Europe*, Franco Angeli Editore, Milano, pp. 271-84

Wynn, M. (1979) 'Barcelona: Planning and Change 1854-1977', *Town Planning Review, 50*, 185-203

— (ed.) (1983) *Planning and Urban Growth in Southern Europe*, Mansell, London

— and Smith, R. (1978) 'Spain: Urban Decentralisation', *Built Environment, 4*, 49-55

Young, G. (1973) *Tourism: Blessing or Blight?*, Penguin, Harmondsworth

Yuill, D., Allen, K. and Hull, C. (1980) *Regional Policy in the European Community: The Role of Regional Incentives*, Croom Helm, London

Zielinski, H. (1983) 'Regional Development and Urban Policy in the Federal Republic of Germany', *International Journal of Urban and Regional Research, 7*, 72-92

Zolberg, A.R. (1974) 'The Making of Flemings and Walloons: Belgium 1830-1914', *Journal of Interdisciplinary History, 5*, 179-235

INDEX